BestMasters

Springer awards „BestMasters" to the best master's theses which have been completed at renowned universities in Germany, Austria, and Switzerland.

The studies received highest marks and were recommended for publication by supervisors. They address current issues from various fields of research in natural sciences, psychology, technology, and economics.

The series addresses practitioners as well as scientists and, in particular, offers guidance for early stage researchers.

Johanna Maria Ticar

3D Analysis of the Myocardial Microstructure

Determination of Fiber and Sheet Orientations

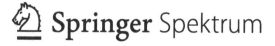

Springer Spektrum

Johanna Maria Ticar
Graz, Austria

OnlinePlus material to this book can be available on
http://www.springer-spektrum.de/978-3-658-11423-7

BestMasters
ISBN 978-3-658-11423-7   ISBN 978-3-658-11424-4 (eBook)
DOI 10.1007/978-3-658-11424-4

Library of Congress Control Number: 2015950752

Springer Spektrum
© Springer Fachmedien Wiesbaden 2016

Printed on acid-free paper

Springer Spektrum is a brand of Springer Fachmedien Wiesbaden
Springer Fachmedien Wiesbaden is part of Springer Science+Business Media
(www.springer.com)

Acknowledgment

This master thesis has been carried out at the Institute of Biomechanics at Graz, University of Technology. I would like to express my gratitude to all people who supported me. Special thanks are mainly due to my supervisor Gerhard Sommer, PhD, for his guidance, contribution of knowledge and experience, support and valuable comments and to the head of the Institute of Biomechanics Prof. Gerhard Holzapfel, PhD. I would also like to convey my gratitude to Andreas Schriefl, PhD, for his effort and valuable help during the data analysis, and Heimo Wolinski, PhD, for his expertise concerning microscopy. Moreover, I am thankful to Dipl.-Ing. Clemens Diwoky, who enabled performing the feasibility study using DTMRI.

Last but not the least important I owe my thanks to my family and friends who supported and encouraged me throughout my years at university.

Contents

List of Figures

4 Discussion 61

Original figures in colour freely available to download from our website www.springer.com

Abstract

Mapping the structure of myocardial fibers is fundamental for assessing the mechanical and electrical properties of the cardiac muscle and hence, the understanding of remodelling following myocardial infarction and other diseases.

Several limitations come along with conventional methods such as histological sectioning and diffusion tensor imaging, which can be eluded by merging multi-photon microscopy with optical tissue clearing. This novel approach yields three-dimensional image stacks featuring the muscle fiber organization and allows the determination of the fiber orientation and its dispersion based on Fourier-based image analysis.

Forty distinct samples from aged but healthy human hearts were used to identify the change of fiber orientations through the ventricular wall. Including 29 samples of the left ventricle wall resulted in a fiber rotation of 196.2 $\pm$ 73.2° (mean $\pm$ standard deviation), that is close to the data reported by previous studies. Furthermore, the respective dispersion parameter yielded 0.16 $\pm$ 0.06, in average. 3-D models were generated by means of volume rendering. They illustrate that cardiac muscle fibers are straight, running in parallel with one preferred fiber direction, however, deposits such as fat seem to compromise the regular and compact structure.

The obtained results show that second harmonic generation imaging combined with optical tissue clearing is an accurate method for determining the three-dimensional muscle fiber orientations and reveals high-resolution insights into the myocardial microstructure. For the very first time, the dispersion parameter of the human myocardium, which can be further used for computational modelling, was determined.

Kurzfassung

Die Darstellung der Struktur der Myokardfasern ist von Bedeutung, um die mechanischen und elektrischen Eigenschaften des Herzmuskels zu bewerten und in der Folge die Remodellierung nach Herzinfarkt oder anderen Erkrankungen zu verstehen.

Konventionelle Methoden wie histologische Schnitte und Diffusionstensor-MRT bergen Nachteile, welche durch den Einsatz von Second-Harmonic-Generation Mikroskopie in Verbindung mit optischem Gewebe-Clearing ausgeschaltet werden können. Dieser neue Zugang liefert dreidimensionale Bildstapel, welche die Ausrichtung der Herzmuskelfasern widerspiegeln und mithilfe von Fourier-basierter Bildanalyse die Bestimmung der Faserausrichtung und deren Dispersion ermöglichen.

Um die Veränderung der Faserorientierung durch die Ventrikelwand zu bestimmen, wurden 40 Proben von gesunden menschlichen Herzen untersucht. Die Analyse von 29 Proben der linken Ventrikelwand ergab eine Faserrotation von $196.2 \pm 73.2°$ (Mittelwert $\pm$ Standardabweichung), ein Wert, der den Ergebnissen früherer Studien entspricht. Der zugehörige Dispersionsparameter weist einen durchschnittlichen Wert von 0.16 ± 0.06 auf. 3-D Modelle, die mittels Volume-Rendering erstellt wurden, verdeutlichen, dass Herzmuskelfasern gerade sind und parallel verlaufen, wobei es eine bevorzugte Ausrichtung gibt. Fettablagerungen scheinen diese regelmäßige und kompakte Struktur zu stören.

Die erhaltenen Ergebnisse zeigen, dass Second-Harmonic-Generation Mikroskopie in Verbindung mit optischem Gewebe-Clearing eine Methode ist, welche die dreidimensionale Muskelfaserausrichtung genau bestimmen lässt und hochauflösende Erkenntnisse zur Mikrostruktur des Myokards liefert. Zum allerersten Mal wurde der Dispersionsparameter des menschlichen Myokards, welcher bei numerischen Modellen und Computersimulationen eingesetzt werden kann, errechnet.

1 Introduction

Motivation

The human heart is composed of a helical network of myocardial fibers, the structure of which is important considering the mechanical and electrical properties of the cardiac muscle (Rohmer *et al.* , 2007). In order to model these properties, the configuration of the 3-D fiber structure has to be understood as modifications in this configuration may cohere with the understanding of remodelling after myocardial infarction (Rohmer *et al.* , 2006). Apart from ischemic heart disease the fiber structure is known to show some modifications in other diseased states, such as ventricular hypertrophy (Smith *et al.* , 2008). Furthermore, the fiber organisation plays a crucial role in arrythmogenesis (Fenton & Karma, 1998), is important considering the activation sequence (Punske *et al.* , 2005), and affects the generation of myocardial stress and strain (Smith *et al.* , 2008). Thus, it is inevitable to know not only the mechanical properties but also the biological microstructure of the heart tissue to develop numerical models and create simulations (Holzapfel & Ogden, 2009). The heart muscle exhibits an anisotropic microstructure implying distinct properties in all directions. As the distortion of the fibers leads to modifications of the electrical and mechanical behaviour, the inherent function of the heart is affected. Hence, the behaviour can be deduced from the microstructure of the heart allowing the identification of diseases such as myocardial infarction amongst others by numerical models.

Structure and properties of the myocardium

The human heart is a muscular vital organ pumping blood through the body. It is divided into four chambers, namely the right and left atria and the right and left ventricles. The blood flows through the heart in just one direction as the atria receive blood from the body and the ventricles pump it around the human body. The volume of the left ventricle is larger than that of the other chambers and so is its wall thickness due to the higher pressure it has to sustain. The wall thickness of the left ventricle alters spatially (Holzapfel & Ogden, 2009); while the wall is thickest at the cardiac base, it gradually gets thinner towards its apex. The common wall thickness of the left ventricle of a healthy human heart is roughly one cm which becomes only one to two mm thick at the very tip of the ventricle (see Fig. 1.1) (Ho, 2009). The wall thickness does not only vary spatially but also temporally along with the cardiac cycle (Holzapfel & Ogden, 2009). The heart wall is located in

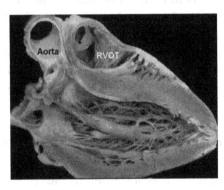

Figure 1.1: Depiction of one half of the left ventricle after sectioning it longitudinally (RVOT = right ventricular outflow tract) (Ho, 2009). (Original figure in colour freely available to download from our website www.springer.com)

the pericardium, a double-walled sac the function of which is to protect the heart. The heart consists of three distinct layers, namely the endocardium (the innermost layer), the myocardium (the middle layer) beyond which is the epicardium (the outer layer) (Ravichandran *et al.* , 2012). The endocardium and epicardium are both membranes with an approximate thickness of about 100 μm consisting mainly of epimysial collagen and elastin. A layer of endothelial cells is another component of the endocardium serving as a common boundary between the blood and the wall (Holzapfel & Ogden, 2009). The myocardium makes up the largest part of the heart wall and is functional tissue as it consists of cardiomyocytes, the contractile cells of the heart, amongst others (Fedak *et al.* , 2005).

The geometry of the left ventricle can be approximated by a thick-walled ellipsoid being truncated at its base. The left ventricular myocardial wall is composed of three-dimensional muscular fibers made of muscle cells, each of which being approximately 80 to 100 μm long. The muscle fibers exhibit a cylindrical shape with a radius of about 5 to 10 μm (Rohmer *et al.* , 2007). Consecutive heart muscle cells are connected via gap (disc) junctions (see Fig. 1.2 (a)) which enable the transmission of electrochemical potentials between them, and hence synchronous beating due to continuous contraction and expansion (Rohmer *et al.* , 2006). The fibers are embedded in an extracellular matrix largely consisting of collagen type I and type III. While collagen type I has a high tensile strength, collagen type III is less rigid and rather contributes to elasticity (Fedak *et al.* , 2005). The fibrillar collagen in the heart can be organised in layers that encompass the myocytes, namely the epimysium, the perimysium and the endomysium. The endomysium, where collagen type III is abundant (Rohmer *et al.* , 2006), surrounds the individual myocytes within each bundle and connects adjacent muscle fibers (Ohayon & Chadwick, 1988), whereas the perimysium connects adjacent muscle layers of a thickness of three or four cells. The perimysium mainly consists of collagen I and the perimysial fibers are oriented parallel to the long axis of the muscle cells (Rohmer *et al.* , 2006). The epimysium ensheats the entire muscle, including both the perimysium and the endomysium. The left ventricular wall is a composite of muscle fibers consisting of myocytes, and these fibers align to form sheets which are physically separated by the perimysium. The muscle fiber orientations change with position throughout the thickness of the wall, as depicted in Fig. 1.3. Figure 1.3 (b) shows the structure through the thickness from the epicardium to the endocardium while (c) shows five sections representing regular intervals from 10 to 90 per cent of the wall thickness. As can be seen in 1.3 (c) (Holzapfel & Ogden, 2009) and according to histological studies and DTMRI measurements (Rohmer *et al.* , 2006), the muscle fiber direction rotates from approximately +60° to -60° across the wall (Rohmer *et al.* , 2007). The fiber angle varies through the thickness of the wall exhibiting nearly 0° in the mid-wall region (Holzapfel & Ogden, 2009).

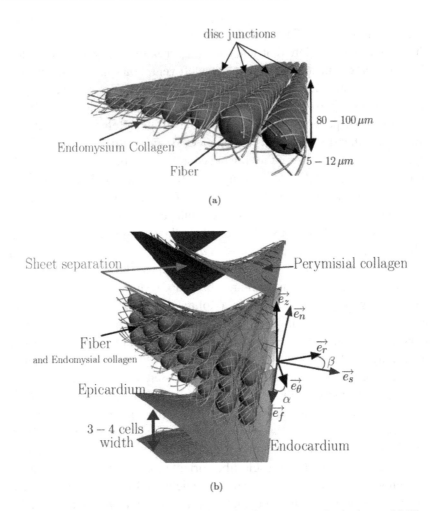

Figure 1.2: Illustration of the fibers and the laminar structure in the heart. (a) The
fibers are represented by the long oval structures; the endomysial collagen, and the disc junctions, connecting individual heart muscle cells,
are shown. (b) Depiction of the laminar structure bounded by cleavage
planes. Fibers align to form sheets grouped in a volume of three to
four cells wide (Rohmer *et al.* , 2006). (Original figure in colour freely
available to download from our website www.springer.com)

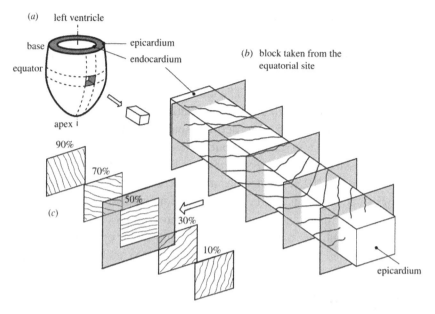

Figure 1.3: (a) Schematic diagram of the left ventricle with a cutout from the equator, (b) of which the structure through the thickness from the epicardium to the endocardium is illustrated. (c) The change in fiber orientation from 10 to 90 per cent of the wall thickness is shown (Holzapfel & Ogden, 2009). (Original figure in colour freely available to download from our website www.springer.com)

Imaging tools for fiber mapping

Several limitations come along with established methods such as diffusion tensor magnetic resonance imaging or histological techniques to determine the fiber structure of the heart. Diffusion tensor imaging requires expensive MRI scanners and is a very time-consuming method. Nonetheless, the spatial resolution is low. Only highly sophisticated MRI scanners allow a suitable resolution of less than 100 μm within a reasonable time span (Smith *et al.* , 2008). Histological sectioning is a difficult and labor-intensive method which may destroy the normal fiber architecture of the biological tissue and it remains a challenge to obtain a three-dimensional visualization of it. An advantage of histological techniques is their capability to achieve much higher spatial resolutions.

The inconveniences that come along with either diffusion tensor imaging or histological sectioning can be avoided by the use of confocal or multi-photon microscopy (Smith *et al.* , 2008). The multi-photon microscopy is attended by low penetration depths that can be dramatically increased using optical clearing (described in 2.2.2) in order to visualize thicker tissue structures (Schriefl *et al.* , 2013). Essentially, optical clearing relies on refractive index matching of tissue light scatterers (such as collagen) and tissue dehydration (Hirshburg *et al.* , 2013). Water has a refractive index distinct from that of biological constituents like collagen, therefore submerging tissue into a clearing agent remains indispensable for improving its penetration depth. Apart from the enhanced penetration depth, non-linear optical techniques using near-infrared wavelengths are favoured due to spatial resolution, excitation capability, and reduced scattering (Cicchi *et al.* , 2013). Multi-photon microscopy, precisely second harmonic generation (SHG), performed on cleared biological tissue allows its three-dimensional visualization, and moreover, avoids manual slicing due to sequential automated slicing of roughly 1.5 mm thick z-stacks (Schriefl *et al.* , 2013).

Due to the long scanning-time and the subsequent low spatial resolution achieved by diffusion tensor imaging while performing a feasibility study, non-linear optical imaging (namely the methodology of second harmonic generation) was chosen over diffusion tensor imaging to determine the muscle fiber and sheet orientation and their dispersion in the human myocardium.

2 Materials and Methods

2.1 Materials

The samples used for the analysis of the myocardial microstructure were provided by the Department of Transplant Surgery, Medical University of Graz, Austria. While a porcine heart (provided by an abattoir) was used for the feasibility study performed by means of diffusion tensor imaging, all further examinations were executed using human heart samples. The different samples were derived from nine distinct patients, both female and male. The information on the patients such as age, sex, size, weight, heart wall thickness, ejection fraction, heart conditions and cardiac diseases, cause of death, and location (left ventricle (LV) vs. right ventricle (RV)) are denoted in Table 2.1. However, we were not endowed with more intrinsic information about heart no. 1, 3, and 9 by the time this thesis was accomplished. Although the 40 samples were derived from 9 different hearts, the data of only 6 patients is provided, resulting in a mean age of 54 years.

Table 2.1: Patient data from the 9 different hearts (the data of heart no. 1, 3, and 9 was not availabe) used, under specification of age, sex, size, weight, wall thickness, ejection fraction, heart conditions and cardiac diseases, cause of death, and location (left ventricle (LV) vs. right ventricle (RV)).

Heart no.	Age (yrs)	Sex	Size (cm)	Weight (kg)	Wall thickness (LVFW/IVW), (mm)	Ejection fraction, (%)	Heart condition/ cardiac diseases	Cause of death	Location	Samples
1	-	-	-	-	-	-	-	-	LV	I
2	57	f	168	63	10/10	58 (56)	slight-moderate MR, HTN	SAH	RV	XXI
3	-	-	-	-	-	-	-	-	LV	II, V, XVI
4	57	f	167	80	11/11	84	concentrial LVH, PAP borderline	aSAH	LV	VI
5	48	m	180	69	9/10	>60	sinus tachycardia, incomplete RBBB, intact LV function, minimal MR,	SAH	LV	III, VII, VIII
6	71	f	169	80	14/14	-	SSS, CAD, paroxymal AF normofrequent after PM implant, HTN	SAH	LV	IV
7	39	m	180	91	-	-	small cardiac and mediastinal silhouette, liver transplant	ICH	LV	XVII, XVIII
8	49	m	180	109	-	-	K1 streptococcus agalactiae, hypernatremia	meningitis, SAH, sepsis	LV	XIX, XX
9	-	-	-	-	-	-	-	-	LV	IX-XV, XXII

AF...atrial fibrillation; aSAH...aneurysmal subarachnoid hemorrhage; CAD...coronary artery disease; HTN...hypertension; ICH...intracranial hemorrhage; IVW...intraventricular wall; LVFW...left ventricular frontal wall; LVH...left ventricular hypertrophy; MR...mitral regurgitation (mitral insufficiency); PAP...pulmonary artery pressure; RBBB...right bundle branch block; SAH...subarachnoid hemorrhage; SSS...sick sinus syndrome

The left ventricular wall is approximately 10−15 mm thick in human adults, however, the exact thickness of all hearts (denoted in Table 2.1) is not given. Moreover, the ejection fraction, the volumetric blood fraction that is pumped out of the ventricle with each heartbeat, of all patients is not available. While an ejection fraction (denoted in %) of more than 60 is considered "nonfailing ", values between 50 and 59 (as heart no. 2) are considered "boarderline ".

2.2 Methods

This chapter describes both imaging methods used, diffusion tensor imaging and multi-photon microscopy, as well as the respective specimen preparation. Furthermore, the software used for data processing and visualization and the accordant method for the analysis of the data are introduced.

2.2.1 Diffusion tensor imaging

Diffusion tensor imaging (DTI) is a magnetic resonance imaging method measuring the rate and direction of the diffusion of water molecules in biological tissues such as the heart. The diffusion tensor obtained from DTI can be decomposed into eigenvalues and eigenvectors of which the principle eigenvector can be used to determine the fiber direction (as diffusion is anisotropic). In case of isotropic environment, the diffusion of water is equivalent in all directions. The water contained inside and ouside the myofibers cannot move freely, but rather in the direction of the fiber itself meaning that the fiber direction corresponds to the maximum direction of diffusivity. The main direction of diffusion is given by the first (and largest) eigenvector. Cleavage planes physically separate the myo- fibers which are grouped in approximately 3-4 cells thick sheets. Diffusion is assumed to be smaller perpendicular to the cleavage planes than inside the cleavage planes. Hence, the third and smallest eigenvector coincides with surface normal direction while the second eigenvector is positioned inside the sheet (Eggen *et al.* , 2012). In order to visualize the anatomic structure of, in this case, the myocardium, the path followed by the fibers has to be reconstructed using a fiber tracking method (Rohmer *et al.* , 2007).

Figure 2.1: Depiction of the fiber structure of the left ventricle following DTI. Fiber tracking was performed and the clockwise to counterclockwise geometry from the epi- to the endocardium is illustrated (Rohmer *et al.* , 2006). (Original figure in colour freely available to download from our website www.springer.com)

Specimen Preparation

As DTI of microstructures is a nontrivial task, the preparation of the samples affects the resulting data. Two principal criteria for DTI are the reduction of water around the samples and the bubble-free preparation of the samples (Rohmer *et al.* , 2006). For the feasibility study a whole porcine heart, from which rectangular samples of approximately $20 \times 10 \times 1.5$ mm size were cut, was used. Immediately after cutting the sample from the heart it was positioned into a container filled with 0.9% PBS solution. The container was put into an ultrasonic bath and sonicated for an hour to eliminate air bubbles. Spherical objects are the most suitable ones for DTI as it is essential to avoid B_0 inhomogenities. Ideally, B_0 fields are homogeneous. In order to eliminate external field inhomogeneities, which would lead to spatial or intensity distortions, the spins have to be excited equally. The bottom of a spherical container was poured with agarose gel, the sample was positioned onto the gel and covered.

Diffusion tensor imaging and processing

The diffusion-weighted MRI data were acquired at the TU Graz MRI laboratory of the LKH Graz and the head coil on a SIEMENS MAGNETOM TrioTim 3T was used. The acquisition was performed over 10 hours of imaging time and 319 images were obtained. The following imaging parameters were used: slice thickness = 5, repetition time = 7.3, echo time = 3.41. The achieved dataset was stored as DICOM diffusion-weighted image metadata.

Data Processing and Visualization

BioimageSuite and 3Dslicer, which are both open source software, were used to visualize and process the data.

BioimageSuite BioimageSuite is an integrated image analysis software suite developed at Yale University supported from the National Institutes of Health (NIH) and the National Institute of Biomedical Imaging and Bioengineering (NIBIB), having its main focus on neuro- and cardiac imaging and analysis. The data obtained from DTI had to be converted into NIFTI format before loading it into the program as BioImageSuite can only process data in ANALYZE or NIFTI format. This was realized using MRIcron. In order to get reasonable results, the diffusion-weighted images were preprocessed using an image filter. Therefore, after loading the images into the program they were filtered by a special diffusion filter termed *Anisotropic Diffusion* aiming to reduce noise while preserving significant parts of the image content. The creation of a mask of the region of interest is another prerequisite before building the diffusion tensor. The Diffusion tensor itself was built with the *Diffusion Tool* of BioImageSuite. BioImageSuite is not capable of recognizing the gradient directions of the dcm2nii conversion files (MRIcron), and hence, they were added manually by means of a MATLAB script as the gradient directions are stored in the DICOM header of each file. From the previous computed diffusion tensor image the diffusion tensor can be further analyzed in order to display characteristics such as the eigenvalues and -vectors, the fractional anisotropy, the directionality, and fiber and sheet angle among others.

3Dslicer   3D slicer is a software package for visualization and image analysis developed by several institutions including the Massachusetts Institute of Technology, the Surgical Planing Laboratory and the National Institute of Health (NIH). The DICOM data obtained from DTI had to be converted to a NRRD format using a NRRD converter before loading it into Slicer for the analysis. The data was preprocessed, first by using a noise reducing filter on the diffusion-weighted imaging dataset, secondly by creating a mask. After preprocessing a diffusion tensor image (coloured by orientation mode) was estimated from the diffusion-weighted image. The images can also be coloured according to their fractional anisotropy. Fractional anisotropy describes the degree of anisotropic diffusion which is zero (dark) for fully isotropic materials and one (bright) for fully anisotropic materials (diffusion along one main axis). DTI tractography, the virtual reconstruction of the trajectory of water molecules along fibers, was performed. However, 3DSlicer is not endowed with a tool to determine the fiber or sheet angle.

2.2.2 Multi-photon microscopy

Multi-photon microscopy (MPM) based on the excitation of fluorescent probes is a standard imaging method (Williams *et al.* , 2005). However, second harmonic generation (SHG) microscopy, first demonstrated in crystalline quartz in 1962 (Campagnola *et al.* , 2002), has been emerging as a powerful imaging tool for biological and biomedical applications, especially for collagen rich tissues such as cornea, tendons and arteries (Cicchi *et al.* , 2013). In contrast to fluorescence microscopy, SHG is a label-free method that does not have the side-effects of photobleaching or toxicity, but still provides high-resolution, high-contrast, and three-dimensional studies of tissues (Campagnola *et al.* , 2002). In contrast to two-photon fluorescence, in second-harmonic imaging, no energy is lost (Cox & Kable, 2006) and the scattering of the illuminated light is less due to the relatively long excitation wavelengths used, allowing deeper penetration depths than would be possible with other microscopy techniques. Moreover, SHG is produced in a small volume, because the intensity of the second harmonic signal depends on the square of the intensity of the incident illumination and thus, on $1/r^4$ (with r being the distance from the focal point of the illumination). The potentially increased spatial resolution results from the nonlinear relation between illumination and SHG intensity (Kumfer, 2009).

SHG microscopy is adapted from the nonlinear optical effect known as second harmonic generation or frequency doubling. When two photons, arising from a laser source, interact with a nonlinear material like collagen, a photon exhibiting half the wavelength and therefore twice the energy and twice the frequency of the incident light will be generated. The Jablonski diagram of SHG is demonstrated in Fig. 2.2.

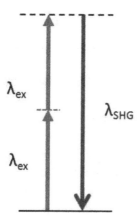

Figure 2.2: Schematic Jablonski diagram of SHG (Larson, 2011). (Original figure in colour freely available to download from our website www. springer.com)

The resulting beam of light, having twice the energy, usually travels in the same direction. As SHG is a coherent process, the resulting beam is also in phase with the exciting light.

In order to generate second harmonics, the electric field of the exciting light has to be strong enough to deform a molecule, and the anisotropy resulting from a non-symmetrical molecule creates an oscillating field at twice the frequency (Cox & Kable, 2006). Hence, the ability to generate second harmonics is peculiar to molecules that lack a center of symmetry. Collagen, being non-centrosymmetric and anisotropic, meets these requirements. Therefore, collagen enclosed structures such as the heart muscle fibers can be visualized by SHG (Campagnola *et al.* , 2002). Moreover, collagen type I is highly crystalline making it even more favourable for SHG, as molecules arranged in a crystalline array generate a much stronger second harmonic response. Even though collagen I and III, both abundant in the ECM surrounding the heart muscle fibers, give a SHG signal, the signal from collagen I is much stronger due to its crystallinity. Besides, SHG imaging allows the differentiation between collagen type I and III by using collagen specific stains. (Cox *et al.* , 2003).

Setup of an multi-photon microscope   Basically, a microscope for SHG imaging consists of a scanning microscope (confocal laser scanning microscope) coupled to a pulsed infrared laser, both mounted to an optical table (Cox & Kable, 2006). Ideally, pulsed lasers with short pulse lengths (femtosecond or picosecond lasers) are used for SHG and are raster-scanned across the sample by means of mirrors to scan the focal point. The short laser pulse lengths enable the high power needed for SHG without destroying the sample. High photon densities are necessary to guarantee the concurrent arrival of two photons in the focal point. In order to focus the laser light on the sample, which may scatter a second harmonic, the laser passes through the objective (Kumfer, 2009). SHG microscopy enables the construction of three-dimensional images of specimen by imaging deeper into the tissues by using near infrared wavelenghts for the incident light beams.

Generally, the resulting beam of light travels in the same direction as the

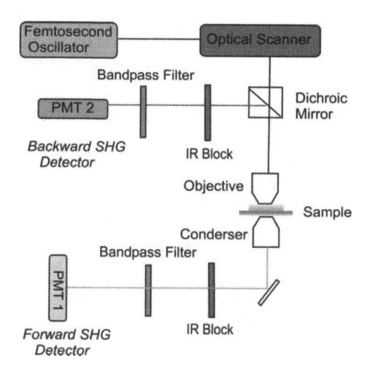

Figure 2.3: Schematic diagram of an SHG microscope (Han *et al.*, 2005). (Original figure in colour freely available to download from our website www. springer.com)

incident beam. However this may alter depending on the illumination conditions, the transparency, and the structure of the specimen. Basically, there is a distinction between forward and backward propagated signals, both the forward and backward SHG detector are depicted in Fig. 2.3. The second harmonic scattered signal is collected by either the objective lense or condenser lense. The immersion objective focuses the exciting beam and collects the backward SHG signals, whereas the condenser collects the forward (transmission) SHG signal through the forward SHG detector. The numerical aperture (NA) of the condenser lens, describing the range within which light can be accepted or emitted, has to be larger than or at least equal to that of the objective (Cox & Kable, 2006). The NA of the objective influences the efficiancy of the backward SHG signals (Han *et al.* , 2005). The amount of room light entering the system is reduced due to the immersion of the system which is an important fact considering nondescanned detectors. However, room light reaching nondescanned detectors should be prevented (Cox & Kable, 2006).

Infrared block filters and bandpass filters in both the forward and backward light paths are in charge of filtering out the illumination light so that only SHG signals are recorded (Han *et al.* , 2005).

Physical background   As electromagnetic radiation (light) passes through a sample (a dielectricum) the electric field E influences the charge distribution, namely separates the positive and negative charges. The subsequent redistribution of charges generates an additional field component, namely the polarization (Cox & Kable, 2006). The total polarization (electric dipole moment per unit volume) of a material interacting with light induced by an electric field E can be expressed as a sum of linear and nonlinear terms, given by

$$P = \chi^{(1)} E^1 + \chi^{(2)} E^2 + \chi^{(3)} E^3 + ... \tag{2.1}$$

where P is the induced polarization, $\chi^{(n)}$ is the n-th order nonlinear susceptibility tensor and E is the electric field. While the first term of the equation describes linear absorption of light, scattering and reflection, SHG is described by the second term of the equation as the nonlinear effect is a polarization of second order in the electric field (Campagnola *et al.* , 2002). These second order effects vanish in materials that have an inversion center (centrosymmetric) or that are isotropic. The nonlinearity between the polarization P and the electric field E in nonlinear media only occurs at very high light intensities induced by pulsed lasers.

Specimen preparation

Cutting the specimen Most of the heart samples of which the microstructure was studied were mechanically tested before, either biaxially or triaxially. Furthermore, in order to determine the fiber orientation throughout the wall, a complete piece of the myocardium was used. All samples were stored fixed in formalin before further use.

Biaxial samples Rectangular samples of approximately $20\times 10\times 2$ mm were cut with a surgical scalpel from formalin fixed specimen previously used for biaxial tensile tests. The longer side of the sample matches the mean fiber direction while the shorter side aligns with the cross-fiber direction as illustrated in Fig. 2.4.

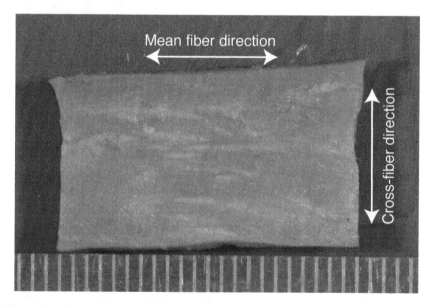

Figure 2.4: Depiction of the fiber vs. cross-fiber direction of a biaxial sample (scale in millimeters). (Original figure in colour freely available to download from our website www.springer.com)

Triaxial samples   There are three different types of triaxial samples, namely NS/NF, SF/SN, and FS/FN, the labeling of which is dependent on the direction they were sheared during the triaxial shear tests. For further information the reader is referred to (Holzapfel & Ogden, 2009).

Left ventricle wall   In order to determine the fiber orientation throughout the entire myo- cardial wall, it was divided into equal slices. This was executed by means of an ordinary meat slicing machine to guarantee that all slices were similarily thick and to obtain straight and smooth surfaces. In order to fix the orientation of the individual slices they were all marked at the same corner as can be seen in Fig. 2.5.

Figure 2.5: Left ventricle wall divided into equal slices. (Original figure in colour freely available to download from our website `www.springer.com`)

Dehydration   The dehydration process (the removal of water) was essential for the further use of benzyl alcohol benzyl benzoate (BABB) as a clearing agent. Therefore, after cutting, the heart samples were dehydrated by a graded ethanol series (50%, 70%, 95%, and twice in 100% for 45 min each). The pure ethanol was thinned out with distilled water. The last step was done twice in undiluted ethanol (100%). Apart from a small deviation, the dehydration was performed according to Smith *et al.* (2008). Instead of 30 minutes cycles, 45 minutes cycles were used, as they showed an improvement of the clearing efficiency of the human myocardium.

Optical photo clearing   After dehydration the samples were treated with BABB (benzyl alcohol:benzyl benzoate) at the ratio of 1:2. The refractive index of BABB is close to that of the heart tissue, and therefore adequate for its optical clearing. Fixed between object holders to avoid possible deformations, the dehydrated samples were stored in the clearing solution and kept in the dark, as BABB is sensitive to sunlight. The samples stayed in the solution at room temperature until imaging. Figure 2.6 displays a sample before and after dehydration and optical clearing.

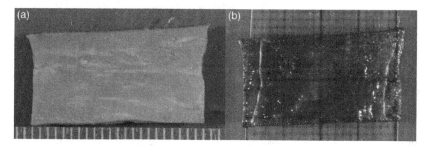

Figure 2.6: Formalin fixed biaxial sample before (a) and after (b) dehydration and clearing (scale in millimeters). (Original figure in colour freely available to download from our website www.springer.com)

Non-linear optical imaging

Figure 2.7: Imaging-setup consisting of a laser (L), the actual confocal microscope (M) and the imaging software (S). (Original figure in colour freely available to download from our website www.springer.com)

SHG-imaging of collagen was performed by means of an imaging-setup consisting of a picosecond laser source and an optical parametric oscillator (OPO; picoEmerald; APE; Germany; HighQ Laser, Austria) integrated into a Leica SP5 confocal microscope (Leica Microsystems, Inc., Austria). To enable the SHG inducing of collagen, the OPO was tuned to 880 nm. For the detection of a backscattered second-harmonic signal, a bandpass (BP) 465/170 emission filter in combination with a nondescanned detector in epi-mode (backward detection) was used. The images were aquired using a Leica HCX IRAPO L 25×0.95 water objective with a working distance (the distance from the front lens of an objective to the focal point) of 2.5 mm for deep tissue imaging (Schriefl *et al.* , 2013). The resulting z-stacks comprised two-dimensional images that were recorded of the previously cleared heart specimen, and each z-stack contains sectional images with a range of vision of 620×620 μm in xy-plane. In depth (z-direction), step sizes were consistent in each recorded stack but varied between 0.4 μm, 5 μm, and 20 μm considering diverse z-stacks. To capture such a z-stack, the top and bottom of the specimen had to be marked and the step sizes had to be defined, as can be seen in Fig. 2.8.

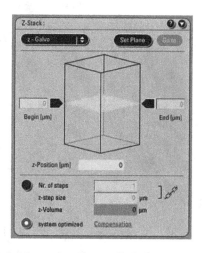

Figure 2.8: Z-stack definitions according to the Leica microscope imaging software. (Original figure in colour freely available to download from our website www.springer.com)

The images were taken with support of Heimo Wolinski, PhD at the Yeast Genetics and Molecular Biology Group, University of Graz, Austria. A representative 2-D image obtained via SHG can be seen in Fig. 2.9.

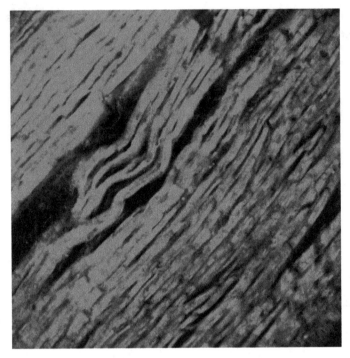

Figure 2.9: Example of a second harmonic generation image of heart muscle fibers with dimensions 620×620 μm. (Original figure in colour freely available to download from our website www.springer.com)

Three-dimensional reconstruction and visualization

Image processing of z-stacks and subsequent volume-rendering were performed using the software package Amira®(Visage Imaging, Inc., Berlin, Germany). Amira is a 3-D visualization and analysis software by the FEI visualization Sciences group for biomedical data obtained from microscopy (amongst other sources).

The data was imported from either a lif-file or multiple tiff-files obtained by SHG imaging. Prior to the actual volume rendering, several different image filters such as Gaussian filtering, unsharp masking, amongst others, were applied to improve the overall outcome.

Data analysis

The recorded z-stacks were first preprocessed using the image processing program ImageJ (ImageJ 1.46r, National Institutes of Health, USA). The actual processing of the image stacks was performed according to Schriefl *et al.* (2013) using MATLAB (Maths Work Inc., MA, USA).
Fiber orientations from two-dimensional images were extracted using a Fourier-based image analysis method in combination with wedge filtering, both prevalent for the characterization of the collagen organization (Schriefl *et al.* , 2012).
Foremost, a window using a raised cosine function was applied to reduce the grey-scale values to zero at the image periphery in order to avoid any frequency-domain effects. A fast Fourier transformation (FFT), represented by the distribution function $f(x, y)$, was applied to the windowed image. A coordinate shift was performed to transform the lowest spatial frequency to the origin. Subsequently, the power spectrum P was calculated by

$$P(u, v) = F(u, v) \cdot F^*(u, v) \tag{2.2}$$

where $F(u, v)$ is the Fourier transform of the function f at point (u, v) and $F^*(u, v)$ is its complex conjugate. To extract the fiber orientations, discriminated by spatial frequency and orientation, wedge-shaped orientation filters were applied. Hence, a discrete distribution of relative amplitudes $I(\Phi)$ (in %) as a function of the respective fiber angle Φ was obtained by summing up every $P(u, v)$ within individual 1° wedges. Due to the shift caused by the FFT, the angular distribution had to be shifted back by 90°. The distribution of fiber angles Φ and amplitudes $I(\Phi)$ was determined for all individual images n in a z-stack, all together representing a three-dimensional dataset per image stack. The three-dimensional distribution of the corresponding amplitudes with 1° resolution (Fig. 2.10) demonstrates the orientation of muscle fibers in a z-stack after having determined the distribution of the fiber angles separately for each image and contains $I(\Phi)$ of each image (Schriefl *et al.* , 2013). The distribution was fitted using Maximum Likelihood Estimation (MLE) and Least Squares Fitting. The amplitudes of the original three-dimensional dataset were smoothed by means of moving average filters in x- and z-directions, respectively. The span range is dependent on the

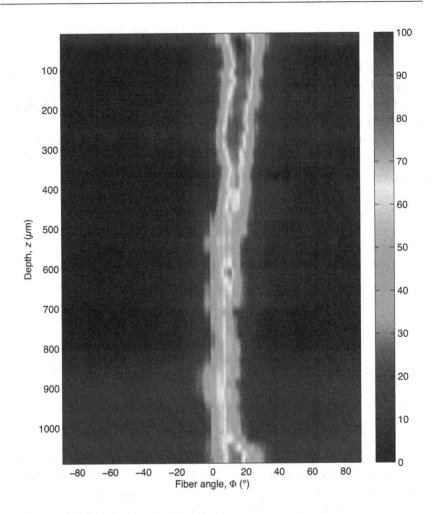

Figure 2.10: Depiction of an intensity plot of a z-stack recorded at a step-size of 20 μm. The different colours ranging from dark blue (0%) to dark red (100%) reflect relative amplitudes. Red areas represent orientations of principal (or preferred) fiber directions, whereas orientations with a low fiber density appear blue. (Original figure in colour freely available to download from our website www.springer.com)

step-sizes in the z-stacks, specified in Table 2.2. Local maxima and minima

Table 2.2: Span range according to x- and z-directions.

Step size, [μm]	x-span	z-span
0.4	9	13
5	9	7
20	9	5

were determined in the smoothed angular distribution yielding the locations of the major peaks. The smallest amplitude value, however, was subtracted from every $I(\Phi)$ in order to remove amplitude offsets. A π-periodic von Mises distribution was used for fitting the angular distribution of fibers, given by

$$p(\Phi) = \frac{\exp(b\cos[2(\Phi - \mu)])}{I_0(b)}. \tag{2.3}$$

with m as the number of peaks (one per image) and $I_0(b_i)$ as the modified Bessel function, given by

$$I_0(b) = \frac{1}{\pi} \int_0^\pi \exp(b\cos\alpha)\mathrm{d}\alpha. \tag{2.4}$$

Φ denotes the angle, while μ and b correspond to the fitting parameters. μ is a location parameter describing the mean (principal) fiber orientation, whereas b is a concentration parameter determining the shape of the von Mises distribution. The b-values, providing information about the density of fibers aligned in the principle fiber orientation μ, refer to a narrower and therefore a more anisotropic distribution when elevated (Schriefl et al. , 2013). Thus, increased values coincide with a low dispersion of the fiber angles and vice versa. Highly aligned fibers are expected from values of $b>1$.

Figure 2.10 depicts the three-dimensional myocytes fiber orientations of a z-stack with a depth of approximately 1060 μm (y-axis). The horizontal lines correspond to about 54 images of which the distribution of fiber angles and amplitudes were indivdually determined.

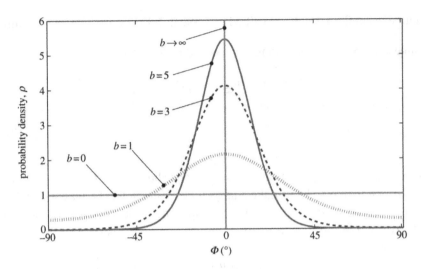

Figure 2.11: Depiction of the von Mises distribution $\rho(\Phi)$ for five different concentration parameters b. $\rho(\Phi)$=1 is true for b=0 characterized by a uniform distribution representing an isotropic fiber distribution. A Dirac delta function at angle μ represents the distribution for $b \longrightarrow \infty$ (Schriefl *et al.* , 2012). (Original figure in colour freely available to download from our website www.springer.com)

The different colours represented in the bar on the right side of the figure correspond to the relative amplitudes of the angles displayed on the x-axis. The colours range from dark blue (0%) to dark red (100%) corresponding to orientations of preferred fiber directions. Hence, while red areas show a high density of fibers in the indicated direction, blue areas display orientations with a low fiber density. A fiber angle of 0° denotes the alignment of fibers in the direction that was previously characterized as fiber direction and thus, corresponds to the horizontal axes of the images.

Additionally, the dispersion parameter κ and R^2, a measure of the goodness of fit, were determined for each distribution. A value of R^2 of one indicates that the data is perfectly fitted, while low values of R^2 refer to a poor regression.

The dispersion parameter κ can be computed from Eq. 2.3, given by

$$\kappa = \frac{1}{\pi} \int\limits_{-\pi/2}^{+\pi/2} \rho(\Phi)\sin^2(\Phi)\mathrm{d}\Phi = \frac{1}{2}(1 - \frac{I_1(b)}{I_0(b)}) \qquad (2.5)$$

with $\kappa \in [0,1/3]$. If there is no dispersion, κ is zero, thereby describing the ideal alignment of fibers, which is represented by $b \longrightarrow \infty$. In case of isotropic fiber distributions κ is 1/3 ($b=0$) (Schriefl *et al.*, 2012).

3 Results

3.1 Diffusion tensor imaging

The feasibility study performed on a small part of a porcine heart wall led
to the following results. Figure 3.1 depicts the fibers of the porcine heart
sample in a variety of views, namely coronal (a), only the coronal plane is
visible (b), sagittal (c), and axial (d) plane. These images were generated
using the fiber tractography tool of the software 3Dslicer (3Dslicer 4.1, MA
Institute of Technology, USA). Figure 3.1 illustrates the direction-encoded

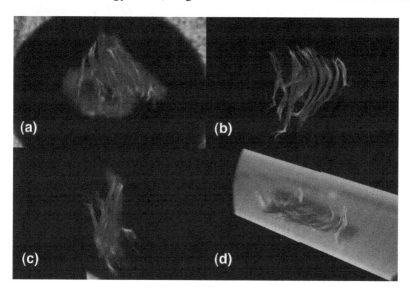

Figure 3.1: Visualization of a porcine heart sample following DTI, depicted in
(a) coronal plane, (b) only the coronal plane is visible, (c) sagittal
plane, and (d) axial plane. (Original figure in colour freely available to
download from our website www.springer.com)

fibers of the heart sample. According to the position of the sample in the MRI scanner, red indicates fibers running in the cross-fiber direction, green in the depth- (z-) direction, while blue coloured fibers run in the mean-fiber direction (fiber direction definitions see 2.2.2).

3.2 Multi-photon microscopy

3.2.1 Three-dimensional reconstruction and visualization

In this section the image stacks (z-stacks) obtained via second harmonic generation (SHG) are shown as volume-rendered 3-D projections. The image stacks were volume-rendered using the software Amira, and feature the heart muscle fibers of distinct samples. The respective three-dimensional reconstructions of the image stacks vary in depth and are displayed from diverging perspectives.

Figure 3.2 represents an image stack (the step sizes between individual images were 5 μm) of heart muscle fibers with a depth of 985 μm, while the in-plane (xy) dimensions are 620×620 μm. The heart muscle fibers are compact and densely packed. These straight fibers that run parallel align to form sheets. In this 3-D reconstruction individual sheets several fibers thick can be observed (note the dark discontinuities appearing in the xz-plane dividing individual sheets).

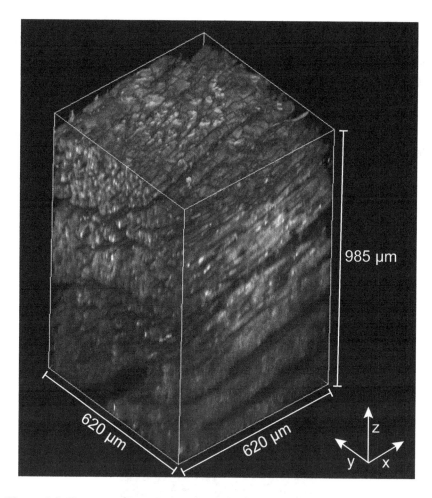

Figure 3.2: Representation of a heart muscle fiber stack volume-rendered using the software package Amira. The overall depth of the depicted z-stack (with a step size of 5 μm between the images) is 985 μm with in-plane dimensions 620×620 μm. (Original figure in colour freely available to download from our website `www.springer.com`)

Figure 3.3 shows 3-D projections of an image stack exhibiting a depth of 640 µm, where (a) reveals a lateral view facing on the xz-plane and (b) facing on the yz-plane. The heart muscle fibers run parallel but seem to be interjected by a fat deposit as can be seen in the xy-plane in (a). This deposit embedded in the heart muscle fibers modifies the fiber orientation. While (a) demonstrates the deposit, the z-stack in (b) which was rotated approximately 90° in order to display the yz-plane, additionally visualizes the formation of cleavage planes.

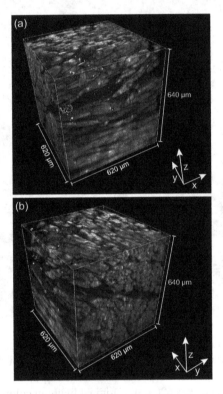

Figure 3.3: Volume rendering of optical sections obtained via SHG imaging. (a) and (b) show the same z-stack featuring a depth of 640 µm, but from distinct perspectives. (Original figure in colour freely available to download from our website www.springer.com)

The z-stack shown in Fig. 3.4 was opened to show the details of the muscle fibers on two opposing surfaces. Therefore, the upper part having a depth of 240 μm was propped up while the lower part exhibiting a depth of 100 μm was folded down. Between the upper and lower part several slices corresponding to a depth of 160 μm were cropped to visualize the change of cardiac muscle fiber organization among a heart sample. Due to the missing slices in the middle corresponding to 160 μm, one can observe the fiber orientation at a depth of 240 μm compared to the fibers at a depth of 400 μm (the in-plane of the lower part). Note the different fiber orientations between the sample shown in Fig. 3.3 where the fibers are orientated along the x-axis, while the sample of Fig. 3.4 exhibits fibers running diagonally.

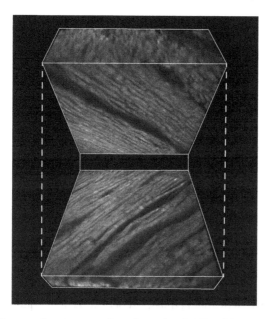

Figure 3.4: This z-stack was opened to show the details of heart muscle fibers on two opposing surfaces. The upper part exhibits a depth of 240 μm, the lower part a depth of 100 μm, while in the middle 160 μm were cropped, leading to a total depth of 500 μm with in-plane dimensions 620×620 μm. (Original figure in colour freely available to download from our website www.springer.com)

While the 3-D projections depicted above all had in-plane dimensions 620×620 µm, the in-plane surface of Fig. 3.5 was cropped to 330×290 µm in x- and y-direction, respectively. The cropping of the in-plane dimensions was necessary due to the computational burden that coincides with the enormous number of individual slices obtained via SHG, as the step sizes between the single images were 0.4 µm and consistent within the z-stack. Hence, 2000 images were used in order to volume-render a sample of a depth of 800 µm. The cardiac muscle fibers are parallel and densely packed. Cleavage planes that separate individual sheets are visible and indicated by the arrows. According to literature (e.g. (Rohmer *et al.* , 2006)), sheets are 3-4 cells wide, while the sheets observable in this 3-D reconstruction seem to exceed this width.

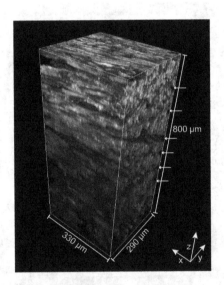

Figure 3.5: Depiction of volume-rendered optical sections aquired by SHG imaging. The in depth (z-direction) step size was 0.4 µm leading to a
total depth of 800 µm. Due to the computational burden, the image
stack was cropped in x- and y-direction to 330 and 290 µm, respectively. The arrows indicate cleavage planes that separate sheets. (Original figure in colour freely available to download from our website
www.springer.com)

The sample illustrated in Fig. 3.6 was rotated around the x-axis before imaging in order to reveal the fact that not many fibers run in the actual z-direction, but are rather organized in sheets. The in-plane (xz) projection of the rotated sample is depicted in (a), while (b) shows a stack of 660 μm depth with the slice distance being 0.4 μm. In both (a) and (b) the organization of fibers in sheets is visible.

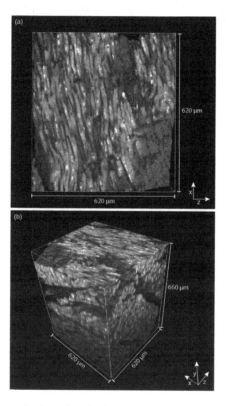

Figure 3.6: Volume rendering of optical sections obtained by SHG imaging. Previous to the actual imaging the sample was rotated 90° around the x-axis. (a) is an in-plane (xz) projection of the top of the volume-rendered sample. (b) shows the 3-D reconstruction of the image stack featuring a depth of 660 μm.(Original figure in colour freely available to download from our website www.springer.com)

Figure 3.7 depicts two different samples, both having a depth of 500 µm with a step size of 5 µm in between individual slices and in-plane dimension 620×620 µm. While the reconstructed sample in (a) exhibits a concentration parameter b of 5.37 and a dispersion parameter κ of 0.10, the concentration parameter b of the sample shown in (b) is 2.40 and hence, less, while the dispersion parameter (κ=0.14) is higher. The data of the stack in (b) is not specified in Table 3.1 as this sample was considered to be an outlier. Although both samples have b-values above 1 (indicating highly aligned fibers), the fiber alignment of the sample shown in (a) is superior. The comparatively increased dispersion parameter κ of the image stack in (b) indicates a more isotropic fiber distribution. One can observe the better arrangment of fibers in (a) as the muscle fibers run more straight and parallel than in (b). Although both samples show the alignment of fibers into sheets, the individual sheets are better observable in (a).

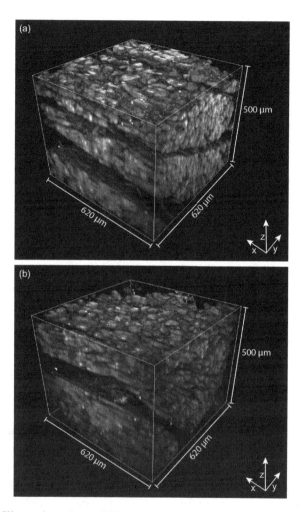

Figure 3.7: Illustration of two different heart samples, both having a depth of 500 μm with a slice distance of 5 μm between indivdual images and in-plane dimensions 620×620 μm. These samples are represented from the same perspective, but the determined concentration parameter b and the dispersion parameter κ are distinct resulting in diverging fiber distributions. (Original figure in colour freely available to download from our website www.springer.com)

3.2.2 Data analysis

The three-dimensional distribution of amplitudes represents the orientation of respective cardiac muscle fibers in a z-stack. The distribution was fitted using Least Squares Fitting and a π-periodic von Mises distribution. Figure 3.8 is a representation of the three-dimensional muscle fiber orientations. The colours ranging from dark blue (0%) to dark red (100%) reflect relative changes in amplitudes. While preferred fiber directions appear red, orientations with a low fiber density appear blue. Hence, blue areas indicate the presence of only few to no fibers. $\Phi = 0°$ denotes the before defined fiber direction corresponding to horizontal aligned fibers (definition of the mean fiber direction, see 2.2.2).

From the fit, both the location parameter μ and the concentration parameter b were obtained and calculated for every single image in the z-stack. The location parameter μ describes the principal (preferred) fiber orientation of the distribution, while the concentration parameter b determines the shape of the von Mises distribution. A narrower distribution indicates elevated b-values and hence, more aligned fibers. Therefore, a more anisotropic fiber distribution follows. Both structural parameters are plotted over the depth, shown in (a) and (b) of Fig. 3.9. The mean fiber orientation, the location parameter μ (disregarding the performed offset correction), plotted over the depth corresponds to the most red parts in the intensity plot Fig. 3.8. The concentration parameter b provides information about the density of fibers aligned in the prinicpal fiber orientation. High values indicate a low dispersion, whereas low values indicate a high dispersion. Higher b-values of the observed specimen demonstrate a more anisotropic cardiac muscle fiber distribution, while a concentration parameter of $b > 1$ already correlates with images of highly aligned fibers. The corresponding coefficient of determination R^2, a measure of the goodness of fit, was determined for each image and plotted over the depth. While a value of $R^2 = 1$ indicates perfectly fitted data, R^2 values that are close to one are satisfactory. Figure 3.9 (d) depicts the disperion parameter κ plotted over the respective depth. The dispersion parameter κ lies in a range between 0 and 1/3 depending on the alignment of the observed muscle fibers, whereupon $\kappa \longrightarrow 0$ in the case of perfectly aligned fibers and a purely anisotropic fiber distribution.

The higher the value of κ, the greater the dispersion and the more isotropic the fiber distribution, with κ being 1/3 indicating complete isotropy (Gasser *et al.* , 2006).

Table 3.1 shows the results from the heart samples of the left ventricle that were either biaxially or triaxially tested before observing the microstructure. The table includes the original position of the samples in the heart (specified by left ventricle (LV), endocardial (end), medial (med) and epicardial (epi)), the depth of the respective heart samples, as well as the computed rotation Φ per depth. The concentration parameter b, the coefficient of determination R^2 and the dispersion parameter κ were calculated for each image in the z-stack, the mean values of which are listed in the table for each of the 29 specimen. For better comparison of the rotation among the different samples, the rotation Φ was determined per mm depth for every sample. Both the mean value and standard deviation (mean $\pm$ SD) were calculated from all the samples listed in Table 3.1 resulting in a rotation of $\bar{\Phi} = 19.62 \pm 7.32°$ per mm depth disregarding possible outliers. In order to spot outliers a outlier detection according to

$$x_i \geq Q_{75} + 1.5 * IQR \text{ or } x_i \leq Q_{25} - 1.5 * IQR \qquad (3.1)$$

with $IQR = Q_{75} - Q_{25} =$ interquartile range, $Q_{25} = 1^{st}$ quartile and $Q_{75} = 3^{rd}$ quartile, was performed. This procedure resulted in one outlier, namely sample III a (shown in Fig. 3.8 and 3.9) exhibiting a rotation of $41.52°$ over one mm depth.

However, due to the fact that the change of orientation is constant throughout the thickness of specimen III a (see intensity plot Fig. 3.8), the mean of the concentration parameter b is 3.70 ± 2.30 (indicating an anisotropic distribution of collagen fibers), and the coefficient of determination R^2 has a rather high mean value of 0.81 ± 0.08, the outlier has not been neglected. The concentration parameter b varied between the different observed specimen, all together resulting in a mean of $\bar{b} = 2.77 \pm 1.53$. The measure of goodness of fit R^2 only exhibits slight distinctions among the probes with $R^2 = 0.83 \pm 0.06$, while the dispersion parameter κ exhibits in average $\bar{\kappa} = 0.16 \pm 0.06$.

The orientation of preferred (or principal) fiber directions appearing red are continuous in the intensity plot of sample VI b (Fig. 3.10), with the area around it being relatively dark blue. Note the bright blue horizontal streak at the depth around 800 µm corresponding to orientations with only few or even no fibers present. The same can be observed regarding the descent of the concentration parameter b in this area which is rather elevated for this sample, exhibiting a mean of 5.10 ± 2.49. Also, the dispersion parameter κ, with a mean of 0.07 ± 0.05, is low indicating great fiber alignment and an anisotropic distribution. However, at the depth of 800 µm the dispersion parameter is highly increased.

While sample III a (Fig. 3.8) and sample VI b (Fig. 3.10) show highly aligned fiber orientations and constant changes in orientation with depth, the regions of preferred fiber directions are not as conspicuous in the intensity plot of sample VIII b (Fig. 3.12) and the change in orientation is slightly varying throughout the thickness of the sample, resulting in a reduced rotation per depth. Note the bright blue streak at a depth of roughly 800 µm, denoting that the fit is not adequate.

Sample XIX b is a triaxial one, the fibers of which are highly anisotropic in the range of 300 to 400 µm (consider the dark blue areas on both sides of the intensity plot and the peak of the concentration parameter b), whereas the anisotropy diminishes after a depth of 400 µm observable at the bright blue or turquoise areas of the intensity plot, the decreasing concentration parameter b, as well as the ascending dispersion parameter κ. However, the coefficient of determination R^2, the measure of goodness of fit, diminishes after a depth of approximately 400 µm.

Table 3.1: Data from 29 distinct heart samples that were previously tested either biaxially or triaxially, under specification of the original position (LV = left ventricle, end = endocardial, med = medial, epi = epicardial; for some samples the exact location was not specified) of the sample in the human heart, the different depths of the samples, the rotation per depth (and per mm depth for better comparison), the concentration parameter b, the coefficient of determination R^2, as well as the dispersion parameter κ. The patient data of the corresponding hearts can be found in Table 2.1 with the same reference number (heart no.).

Sample no.	Stack	Heart no.	Position	Depth, (μm)	Rotation /depth, (°)	Rotation /mm, (°/mm)	Concentration parameter, b	Coefficient of determination, R^2	Dispersion parameter, κ
					Biaxial samples				
I	a	1	LV med	645	6.57	10.19	5.37	0.82	0.10
I	b	1	LV med	600	4.77	7.95	3.90	0.87	0.10
II	a	4	LV med	1280	24.39	19.05	4.21	0.75	0.13
III	a	5	LV epi	840	34.88	41.52	3.70	0.81	0.11
IV	a	6	LV epi	1300	24.66	18.97	1.75	0.76	0.19
V	a	3	LV epi	960	13.15	13.70	3.62	0.89	0.11
V	b	3	LV epi	1060	15.00	14.15	5.30	0.79	0.09
VI	a	4	LV	620	12.73	20.53	4.33	0.74	0.11
VI	b	4	LV	1040	25.46	24.48	5.10	0.79	0.07
VII	a	5	LV end	560	9.03	16.13	4.66	0.80	0.11
VII	b	5	LV end	540	9.33	17.28	2.25	0.85	0.14
VIII	a	5	LV end	760	16.55	21.78	3.86	0.91	0.08
VIII	b	5	LV end	860	9.32	10.84	3.30	0.81	0.10
IX	a	9	LV med	780	10.51	13.47	1.83	0.88	0.17
X	a	9	LV med	660	14.95	22.66	3.11	0.88	0.11
XI	a	9	LV end	660	19.02	28.82	4.60	0.92	0.08
XII	a	9	LV end	600	9.71	16.19	3.58	0.87	0.10
XIII	a	9	LV	740	11.52	15.57	1.30	0.90	0.21
XIV	a	9	LV	620	11.60	18.71	1.15	0.88	0.23
XIV	b	9	LV	580	9.60	16.55	1.10	0.85	0.23
XV	a	9	LV	460	10.50	22.83	1.50	0.79	0.19
					Triaxial samples				
XVI	a	3	LV med	800	26.90	33.63	1.02	0.90	0.23
XVI	b	3	LV med	700	19.65	28.07	0.83	0.91	0.25
XVII	a	7	LV med	680	16.01	23.54	0.70	0.86	0.26
XVIII	a	7	LV med	580	16.75	28.88	0.70	0.82	0.27
XIX	a	8	LV med	1080	17.90	16.57	1.62	0.73	0.20
XIX	b	8	LV med	760	12.54	16.50	2.49	0.76	0.16
XX	a	8	LV med	900	14.60	16.22	1.31	0.75	0.21
XX	b	8	LV med	1000	14.33	14.33	2.13	0.80	0.18

The average depth of these 29 samples is 782 μm. A mean rotation of $\bar{\Phi} = 19.62 \pm 7.32°$ per mm depth was determined. The following average values were calculated: concentration parameter $\bar{b} = 2.77 \pm 1.53$, coefficient of determination $\bar{R}^2 = 0.83 \pm 0.06$, and dispersion parameter $\bar{\kappa} = 0.16 \pm 0.06$.

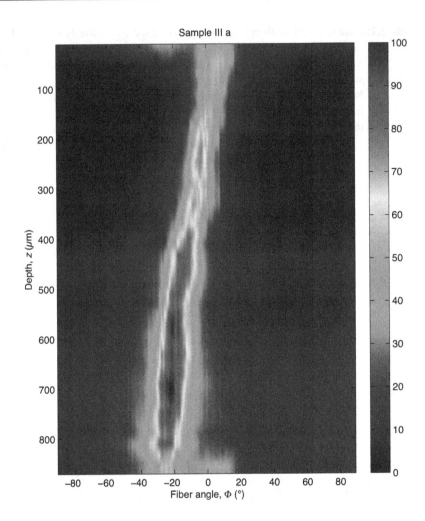

Figure 3.8: Intensity plot of the three-dimensional myocytes fiber orientations of sample III a through a thickness of 840 μm. The colours ranging from dark blue (0%) to dark red (100%) correspond to the relative amplitudes of the angles displayed on the x-axis. Red areas show the preferred fiber orientations and thus, a high density of fibers in the indicated direction, whereas blue areas display orientations with a low fiber density. (Original figure in colour freely available to download from our website www.springer.com)

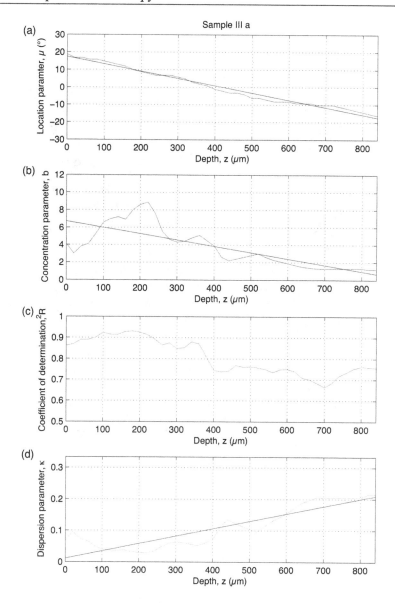

Figure 3.9: Changes of preferred fiber orientations (location parameter μ) and fiber alignment (concentration parameter b) of sample III a over a depth of 840 μm are shown. The coefficient of determination R^2, and the dispersion parameter κ were plotted over the depth, respectively. (Original figure in colour freely available to download from our website www.springer.com)

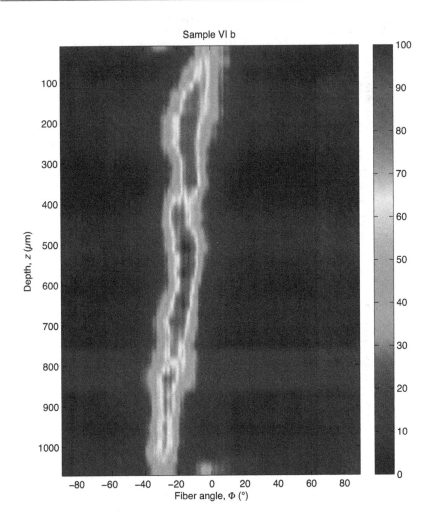

Figure 3.10: Intensity plot of the three-dimensional myocytes fiber orientations of sample VI b through a thickness of 1040 μm. The colours ranging from dark blue (0%) to dark red (100%) correspond to the relative amplitudes of the angles displayed on the x-axis. Red areas show the preferred fiber orientations and thus, a high density of fibers in the indicated direction, whereas blue areas display orientations with a low fiber density. (Original figure in colour freely available to download from our website www.springer.com)

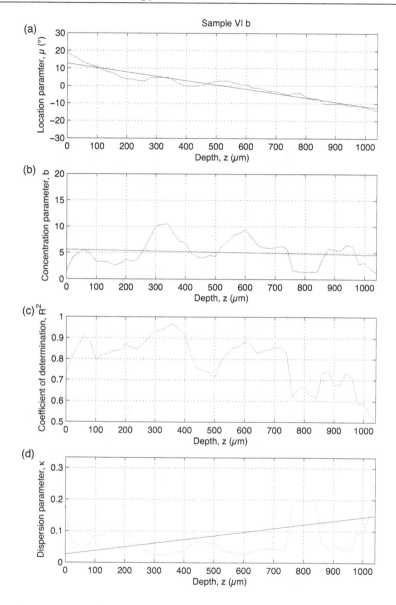

Figure 3.11: Depiction of the preferred fiber orientations (location parameter μ), the fiber alignment (concentration parameter b), the coefficient of determination R^2, and the dispersion parameter κ of sample VI b over the respective depth of 1040 μm. (Original figure in colour freely available to download from our website www.springer.com)

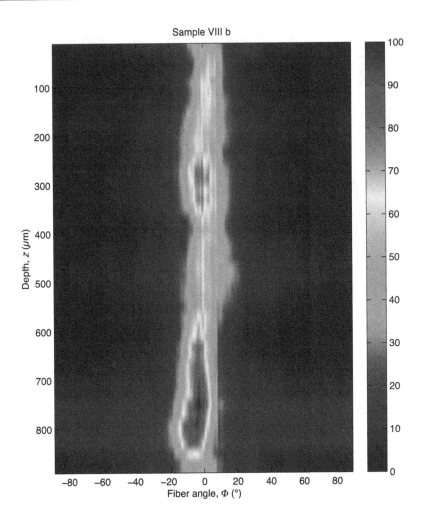

Figure 3.12: Intensity plot of the three-dimensional myocytes fiber orientations of
sample VIII b through a thickness of 860 μm. The colours ranging
from dark blue (0%) to dark red (100%) correspond to the relative
amplitudes of the angles displayed on the x-axis. Red areas show the
preferred fiber orientations and thus, a high density of fibers in the
indicated direction, whereas blue areas display orientations with a low
fiber density. (Original figure in colour freely available to download
from our website www.springer.com)

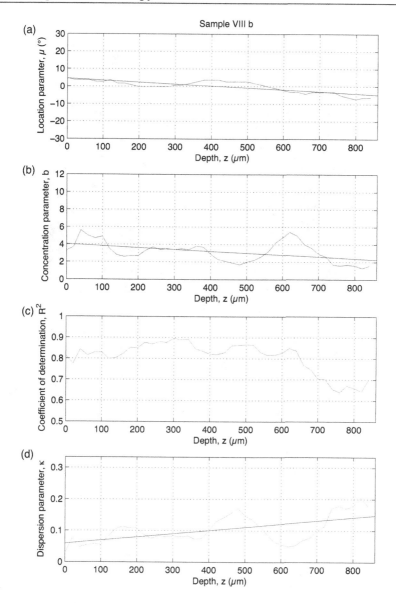

Figure 3.13: Depiction of the preferred fiber orientations (location parameter μ), the fiber alignment (concentration parameter b), the coefficient of determination R^2, and the dispersion parameter κ of sample VIII b over a depth of 860 μm. (Original figure in colour freely available to download from our website www.springer.com)

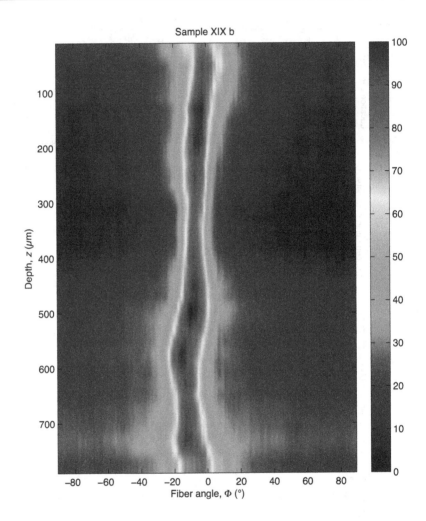

Figure 3.14: Intensity plot of the three-dimensional myocytes fiber orientations of sample XIX b through a thickness of 760 μm. The colours ranging from dark blue (0%) to dark red (100%) correspond to the relative amplitudes of the angles displayed on the x-axis. Red areas show the preferred fiber orientations and thus, a high density of fibers in the indicated direction, whereas blue areas display orientations with a low fiber density. (Original figure in colour freely available to download from our website www.springer.com)

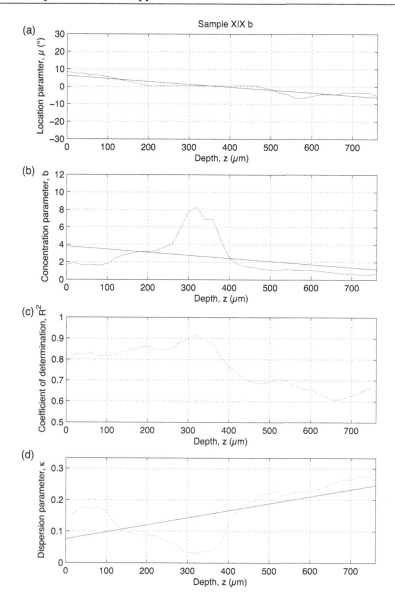

Figure 3.15: Depiction of the preferred fiber orientations (location parameter μ), the fiber alignment (concentration parameter b), the coefficient of determination R^2 and the dispersion parameter κ of sample XIX b over a depth of 760 µm. (Original figure in colour freely available to download from our website www.springer.com)

Although the focus of this work was the determination of the fiber orientation in the left ventricle of the human heart, it was also determined for two samples of the right ventricle, denoted in Table 3.2. As it is apparent from Table 3.2, the rotation per mm depth is higher in the right ventricle than in the left one, although the thickness of the right ventricle wall is less (approximately 4 mm). However, the quantity of samples from the right ventricle tested was minimum, which did not allow for any conclusions to be drawn.

Table 3.2: Data from two heart samples from the right ventricle under specification of the original position (RV = right ventricle, end = endocardial, med = medial, epi = epicardial) of the sample, the respective depth, the rotation per depth and mm depth, the concentration parameter b, the coefficient of determination R^2, and the dispersion parameter κ. The patient data of the corresponding hearts can be found in Table 2.1 with the same reference number (heart no.).

Sample no.	Stack	Heart no.	Position	Depth, (μm)	Rotation /depth, (°)	Rotation /mm, (°)	Concentration parameter, b	Coefficient of determination, R^2	Dispersion parameter, κ
XXI	a	2	RV med	567	20.48	36.11	4.25	0.78	0.10
XXI	b	2	RV med	580	31.99	55.16	2.67	0.72	0.16

Figure 3.16 and Fig. 3.17 depict the intensity plot as well as the location paramter μ, the concentration parameter b, the coefficient of determination R^2, and the dispersion parameter κ plotted over a depth of 567 μm of sample XXI a, the slice distance of which between individual slices is 0.4 μm. The turquoise streak at a depth of 450 μm in the intensity plot can also be observed at the elevated dispersion parameter κ (plotted over the depth in Fig. 3.17 (d)), reaching almost 0.3 indicating isotropic fiber distributions. Almost the same characteristics (a bright blue streak) can be observed at a depth of about 230 μm, where the concentration parameter b is low and the dispersion parameter κ shows a peak beyond 0.2.

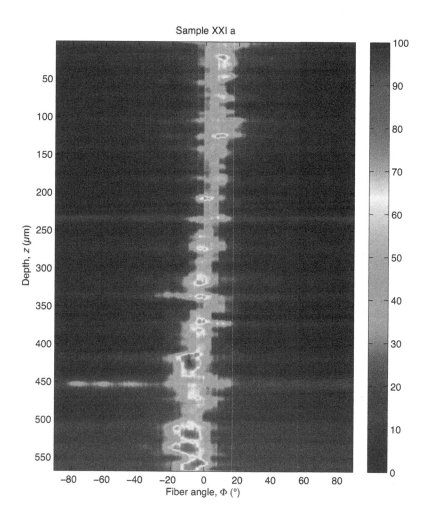

Figure 3.16: Intensity plot of the muscle fiber orientations of sample XXI a derived from the right ventricle over a depth of 567 μm. The colours ranging from dark blue (0%) to dark red (100%) correspond to the relative amplitudes of the angles displayed on the x-axis. Red areas show the preferred fiber orientations and thus, a high density of fibers in the indicated direction, while blue areas display orientations with a low fiber density. The step sizes between individual slices was 0.4 μm. (Original figure in colour freely available to download from our website www.springer.com)

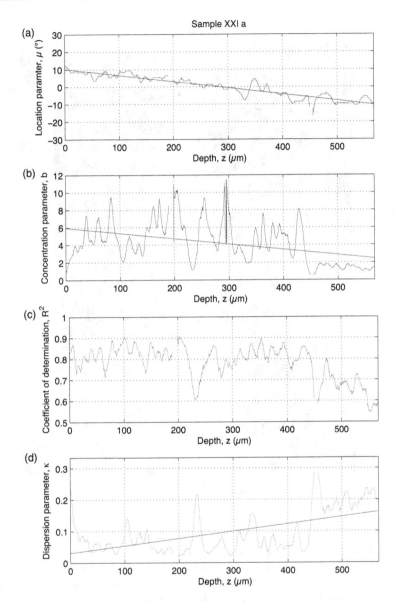

Figure 3.17: Depiction of the preferred fiber orientations (location parameter μ), the fiber alignment (concentration parameter b), the coefficient of determination R^2, and the dispersion parameter κ of sample XXI a over a depth of roughly 567 μm. This sample was derived from the right ventricle. (Original figure in colour freely available to download from our website www.springer.com)

Additionally, to reveal the fact that few fibers are oriented in z-direction, but are rather organized in sheets, a small part of two distinct biaxial heart samples was cut from the middle and rotated 90° around the x-axis for SHG imaging. This rotation was performed to ensure that possible fibers might not get cut by the sectional images making up a z-stack while processing in the original z-direction. After recording and processing the image stack of the rotated specimen, fibers running out-of-plane and fibers oriented in sheets could be identified. The results arising from the stack consisting of xz-planes are shown in Table 3.3. The rotation per mm depth only slightly varies between the two specimen tested, but in contrast to the stacks recorded in xy-plane, the change in the calculated angles is much lower.

Table 3.3: Data from two biaxial heart samples of which a small part was cut out and rotated 90° to visualize the alignment in sheets and to determine the out-of-plane rotation. The original position (LV = left ventricle, end = endocardial, med = medial, epi = epicardial) of the sample, the respective depth, the rotation per depth and mm depth, the concentration parameter b, the coefficient of determination R^2, and the dispersion parameter κ are denoted. The patient data of the corresponding hearts can be found in Table 2.1 with the same reference number (heart no.).

Sample no.	Stack	Heart no.	Position	Depth, (µm)	Rotation /depth, (°)	Rotation /mm, (°)	Concentration parameter, b	Coefficient of determination, R^2	Dispersion parameter, κ
I	c	1	LV med	620	1.54	2.49	4.66	0.79	0.11
I	d	1	LV med	596	1.48	2.48	3.28	0.81	0.12

The corresponding plots of sample I c are shown in Fig. 3.18 and Fig. 3.19. The slight change of orientation throughout the depth of 620 µm can be observed in Fig. 3.19 (a) demonstrating the location parameter μ plotted over the respective depth. While the concentration parameter b decreases with progressive depth, the dispersion parameter increases, both indicating that the dispersion is lower, and thus the anisotropy higher at the beginning. Beyond a depth of 550 µm the isotropy is the highest.

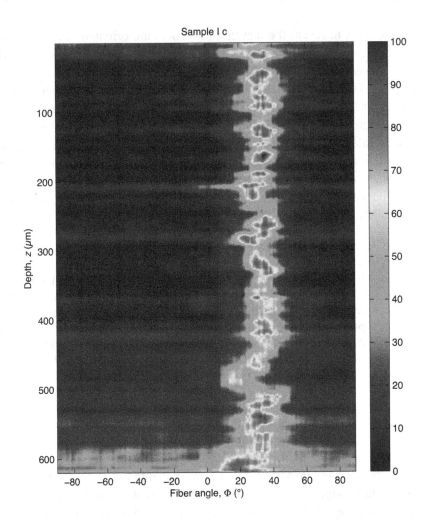

Figure 3.18: Intensity plot of the muscle fiber orientations of sample I c, which was rotated 90° around the x-axis prior to imaging, over a depth of 620 μm. The colours ranging from dark blue (0%) to dark red (100%) correspond to the relative amplitudes of the angles displayed on the x-axis. Red areas show the preferred fiber orientations and thus, a high density of fibers in the indicated direction, while blue areas display orientations with a low fiber density. The slice distance between individual slices is 0.4 μm. (Original figure in colour freely available to download from our website www.springer.com)

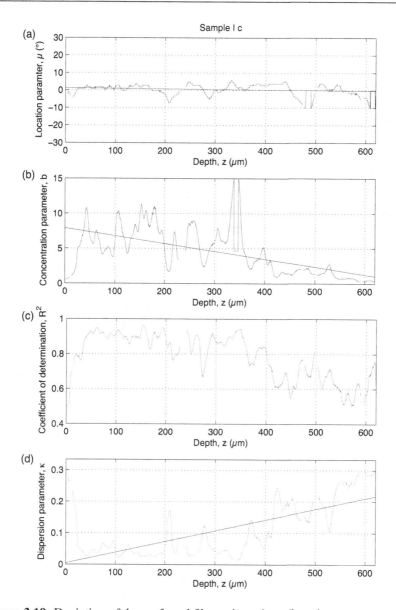

Figure 3.19: Depiction of the preferred fiber orientations (location parameter μ), the fiber alignment (concentration parameter b), the coefficient of determination R^2, and the dispersion parameter κ of sample I c over a depth of 620 μm. This sample was rotated 90° around the x-axis prior to imaging. (Original figure in colour freely available to download from our website www.springer.com)

Fiber orientations throughout the left ventricle wall

For the determination of the fiber orientation throughout the thickness of the left ventricle wall, it was divided into seven equal slices, which were separately imaged and processed. The first slice is closest to the epicardium, while slice seven composes the most endocardial slice of the ventricle wall. The summation of the depths of the slices yields a total thickness of 7580 µm. It has to be considered that a few slices at the very top and bottom of each sample had to be eliminated for efficient data processing and analysis. This slightly diminished the actual depth of each sample about 20 to 80 µm, but improved the outcome, as the resolution and quality of the images at the end of a z-stack are inferior to the ones at the initiation. The slices were imaged at exactly the same spot in order to determine the orientation throughout the thickness of the wall. The results of each slice are denoted in Table 3.4.

Table 3.4: Data from seven slices making up the left ventricle wall under specification of the original position (slice 1 to 7 ranging from the epicardium to the endocardium), the respective depth, the rotation per respective depth and mm depth, the concentration parameter b, the coefficient of determination R^2, and the dispersion parameter κ. The patient data corresponding to these 7 slices constituting the ventricle wall depth are not specified, as they were taken from heart no. 9.

Sample no.	Slice	Position	Depth, (µm)	Rotation /depth, (°)	Rotation /mm, (°)	Concentration parameter, b	Coefficient of determination, R^2	Dispersion parameter, κ
XXII	1	LV epi	860	11.93	13.87	1.00	0.87	0.26
XXII	2	LV epi	1320	13.74	10.41	1.20	0.87	0.23
XXII	3	LV med	980	3.95	4.03	1.23	0.94	0.21
XXII	4	LV med	800	192.11	240.14	1.37	0.83	0.22
XXII	5	LV med	900	21.01	23.34	1.24	0.96	0.22
XXII	6	LV endo	1180	9.47	8.03	0.54	0.84	0.28
XXII	7	LV endo	1540	12.11	7.86	1.01	0.79	0.24

While the rotation of slice three over a depth of 980 µm with 3.95° is comparatively small, slice four exhibits an exceeding change in orientation. This is presumably due to a fat deposit invading the distribution of cardiac muscle fibers and hence, has a large impact on the results. The respective plots of slice four are constituted in Fig. 3.20. Until a depth of 350 µm the fibers show preferably isotropic behaviour (note the low concentration parameter b and the high dispersion parameter κ), however, the goodness of fit R^2 is consistently high. At a depth of approximately 400 μm a sudden change of orientation is visible due to the embedded deposit. At a depth of roughly 500 μm highly aligned fiber orientations can be observed (note the dark blue region next to the red area of preferred fiber orientations in the intensity plot on the left of Fig. 3.20). However, the coefficient of determination R^2 slightly diminishes with progressive depth. After a depth of 700 μm the fiber directions are mainly isotropic dispersed which can be observed in the intensity plot as well as the missing data points on the plots on the right side of the figure. The regions with no data points correspond to areas with almost isotropic fiber distributions (Schriefl *et al.* , 2013). The SHG images of slice four at a depth of (a) 100 μm (b) 200 μm (c) 300 μm (d) 400 μm (e) 500 μm (f) 600 μm are illustrated in Fig. 3.21 for better comprehending the source of the exceptional rotation.

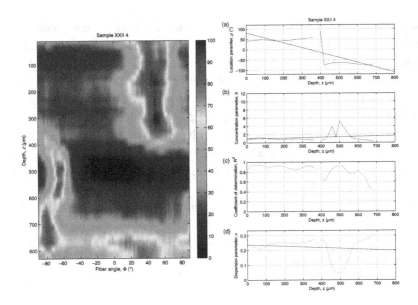

Figure 3.20: Representation of the intensity plot of the fourth slice of sample XXII as well as the location parameter μ, the concentration parameter b, the coefficient of determination R^2, and the dispersion parameter κ plotted over the respective depth. (Original figure in colour freely available to download from our website www.springer.com)

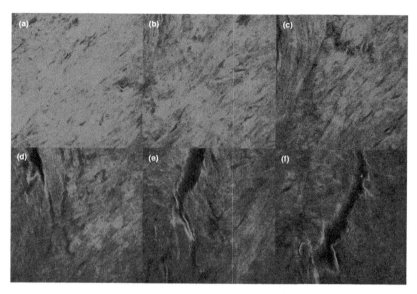

Figure 3.21: Illustration of images obtained via SHG at different depths of probe XXII slice 4 to visualize the deposit embedded in the collagen fibers (a) 100 μm (b) 200 μm (c) 300 μm (d) 400 μm (e) 500 μm and (f) 600 μm. (Original figure in colour freely available to download from our website www.springer.com)

4 Discussion

This chapter covers the conveniences and limitations of the used methods and debates the results.

4.1 Diffusion tensor imaging

Although diffusion tensor magnetic resonance imaging has already been successfully used for the reconstruction and visualization of fiber and laminar structure in the human heart (Rohmer *et al.*, 2007), as well as in that of animals such as mice (Peeters *et al.*, 2006), rats (Chen *et al.*, 2003), (Sosnovik *et al.*, 2009), sheep (Kung *et al.*, 2011), and others, the method of multi-photon microscopy was chosen over diffusion tensor MRI after performing a feasibility study with a porcine heart.

Even though the aim of this work was the determination of the fiber structure of the human heart, a porcine heart was used for the feasibility study because it is more easily obtainable than a human heart and exhibits both structural similarities and characteristics to the human heart. The preparation of the porcine heart probe was carefully executed to avoid the generation of air bubbles around the probe that would affect the precise aquisition by means of diffusion tensor MRI. The probe was put into a PBS (phosphate puffered saline) solution, a buffer solution based on water containing salt, having the advantages of being isotonic and non-toxic to cells and maintaining a constant pH. Immediatly after the insertion of the probe in PBS it was put into an ultrasonic bath to remove inevitable air bubbles. The sonication lasted approximately one hour, however, the results could be improved by sonicating the probe during an even longer time period. Another important prerequisite of the specimen preparation for diffusion tensor MRI is the reduction of the amount of water surrounding the probe.

It was decided to put the rectangular heart probe into a spherical object to preclude B_0 inhomogeneities which would lead to mismapping of tissue as an inhomogeneous external field may either result in spatial or intensity distortions (Erasmus *et al.* , 2004). The spherical container was filled with agarose gel serving as an MRI phantom as it is a magnetic resonance signal bearing material. However, other signal bearing solutions could have been used. After preparing the agarose gel, it was carefully poured into the spherical container to avoid the generation of bubbles which would affect the imaging itself. The pouring of agarose gel was performed instantly after its preparation to ensure its fluidity as it solidifies at room temperature within short time. For this reason the agarose gel was still hot when poured into the container and the heart sample was placed. The heat arising from the agarose gel and percolating the heart sample might influence its microstructure and therefore, the orientation of the myofibers might be slightly modified.

The diffusion tensor imaging itself, performed at the TU Graz MRI laboratoray at the LKH Graz overnight, lasted several hours. The long imaging times, however, led to only insufficient results. It is well known from literature that clinical tomographic modalities such as MRI including diffusion tensor imaging have a limited resolution of roughly one mm (Campagnola, 2011). However, longer imaging periods, e.g. over almost 60 hours of imaging time (as performed by (Rohmer *et al.* , 2007)), would probably improve the outcome. Preclinical imaging, such as micro-MRI, having a substantially improved spatial resolution, is definitely more suitable than normal human MRI scanners to examine the microstructure of the human heart. However, such scanners were not readily available.

The programs BioImageSuite and 3Dslicer were both developed for MRI image processing, however, they exhibit several differences. 3D slicer was preferred to BioImageSuite for the purpose of this work, as it was not possible to get a reasonable illustration of the muscle fibers using the software BioImageSuite without any additional programing. The images obtained via 3Dslicer are depicted in the previous chapter, however, 3Dslicer does not possess a tool for the determination of the fiber and sheet angle. Yet, in our perception, the resolution achieved during the feasibility study was not suitable and therefore, this work was continued using multi-photon microscopy.

4.2 Multi-photon microscopy

As established methods such as diffusion tensor imaging and histological sectioning entail inconveniences, the novel approach of multi-photon microscopy combined with optical tissue clearing was used, allowing a fast and automated analysis of the microstructure of cardiac tissue. The used method enables the representation of heart muscle fibers in a z-stack as a three-dimensional orientational distribution and moreover, the determination of material parameters, which can be further used in numerical modelling and finite element analyses. Forty different tissue samples from healthy human hearts were analyzed to observe their microstructure.

Although the samples were derived from 9 distinct hearts, the data of only 6 patients, including both female and male, is available, resulting in a mean age of 54 years. The ejection fraction of only 3 patients is provided. While 2 of them have an ejection fraction >60%, which is considered "nonfailing ", the ejection fraction of one patient is slightly reduced, and therefore, considered "boarderline ". All patients had mild or modest cardiac diseases, which are denoted in Table 2.1, however, none of these patients died of cardiac failure. The cause of death in most cases was subarachnoid hemorrhage (SAH).

The elevated resolution and specificity of SHG microscopy in contrast to clinical imaging modalities such as diffusion tensor MRI, which image through a considerably higher depth, come at the expense of limited penetration depths of 100 to 300 μm. However, the combination of second harmonic generation with optical tissue clearing eludes this drawback. As can be seen in Fig. 2.6, optical clearing definitely leads to an elevated transparency and hence, the penetration depth is increased severalfold (Campagnola, 2011). Due to the optical tissue clearing performed on the samples beforehand, penetration depths ranging from 600 up to 1500 μm could be achieved by means of SHG. Optical clearing does not affect the fiber orientations of tissue samples, as was shown by Schriefl *et al.* (2013) by comparing the fiber distributions before and after the optical clearing process.

According to literature, namely Smith *et al.* (2008), the dehydration process previous to the optical clearing was performed by a graded ethanol series in 30-minute-cycles for porcine hearts. However, as preliminary tests showed that the dehydration process of the human myocardium improves with a slightly extended duration of the cycles, the samples stayed in the respective dilution grades for 45 minutes. Exposure times longer than 45 minutes, however, did not improve the results any further.

After dehydration, a shrinkage of roughly 10% of the sampes was observed. The shrinkage due to dehydration is consistent with literature ((Cheng *et al.* , 2005), (Smith *et al.* , 2008)). As the measured shrinkage of the samples is rather low and uniform in all directions, it is assumed to not affect the analysis of the structure.

Figure 4.1 demonstrates the difference between a freshly cleared sample and one that stayed in the clearing agent for several months. The sample that was only imaged after staying in the clearing agent for five months exhibits some fractures in the structure and the fibers seem to be broken as depicted on the right side of Fig. 4.1. Therefore, such a long duration of stay in the clearing agent should be avoided and the imaging should be performed soon after the effective clearing process.

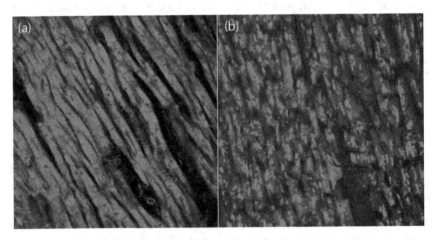

Figure 4.1: Distinction between a recently cleared sample (a) and a sample that stayed in the clearing agent BABB for more than five months (b). (Original figure in colour freely available to download from our website www.springer.com)

Intact imaging via SHG presumes objects to have a smooth surface, however, this condition was not entirely fulfilled due to the manual cutting of the heart samples or possible deformations that occurred during the sample's stay in the clearing agent. Therefore, the field of view was not uniform throughout the sample, but varying at the very end of the sample leading to images seen in Fig. 4.2, demonstrating deficient fields of view. To avoid these irregularities at the surfaces of the samples, some of the most upper images were erased for both the visualization and the calculation. The removal of some images at the beginning of the recorded z-stacks diminished the actually achieved penetration depth, however, the overall outcome was improved.

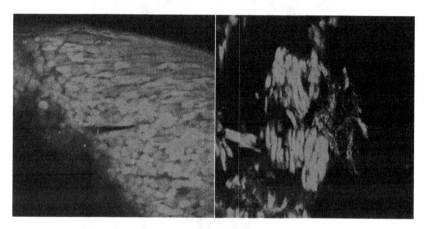

Figure 4.2: Examples of deficient field of views due to the manual cutting of the heart samples. (Original figure in colour freely available to download from our website www.springer.com)

In some of the images black spots (as depicted in Fig. 4.3 (a) are visible. These are fat deposits invading the microstructure of the heart. Also the deposit shown in (b) disturbs the usually straight and parallel running heart muscle fibers.

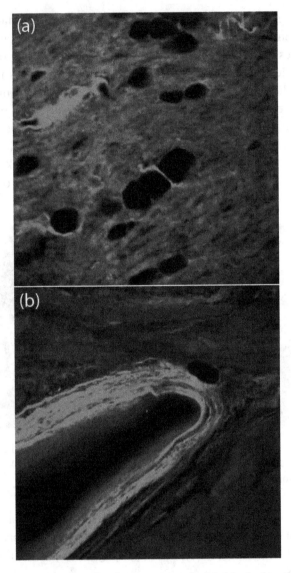

Figure 4.3: SHG images showing fat deposits embedded in the heart muscle fibers. (Original figure in colour freely available to download from our website www.springer.com)

Figure 4.4 depicts an image of the heart muscle fibers obtained via SHG that was detected in forward direction instead of the normally used epi-mode, thus demonstrating that backward propagating SHG led to high-resolution images while forward propagating SHG came along with noise.

Figure 4.4: Depiction of an SHG image of cardiac muscle fibers detected in forward direction. (Original figure in colour freely available to download from our website www.springer.com)

While scanning by means of an SHG microscope it is advisable to darken the room lighting and optically isolate or cover the microscope, elsewise room light will be detected hindering the effectual experiment by adding noise. This is important to consider using any form of nondescanned detector (Cox & Kable, 2006).

Volume-rendering enabled the generation of 3-D models of the image stacks obtained by SHG. Several basic similarities are visible in the 3-D reconstructions shown in section 3.2.1. It is obvious that cardiac muscle fibers are straight, running in parallel with one preferred fiber direction, however, deposits such as fat seem to compromise the regular and compact structure. The 3-D reconstructions enable the visualization of the alignment of fibers in sheets, which, according to literature (Rohmer *et al.* , 2007), are 3−4 cells wide. Nevertheless, the sheets observable in the 3-D reconstructions (see 3.2.1) seem to exceed this width, as they are grouped in a volume of approximately 4−10 cells wide. Visually, it is not possible to distinguish the samples according to their location within the heart due to their epicardial, medial or endocardial origin.

To extract fiber orientations from individual SHG images a Fourier-based image analysis in combination with wedge filtering was used (Schriefl *et al.* , 2013). One principal (or preferred) direction of fiber orientation refers to anisotropic regions for the distribution fitting of which a von Mises distribution was used. Hence, the principal fiber orientation μ and the concentration parameter b, a measure of the fiber density around μ, could be determined, from which the dispersion parameter κ was calculated accordingly. These parameters can be further used to develop numerical models and to create simulations. Distribution fitting was performed through both Maximum Likelihood Estimation (MLE) and Least Squares Fitting, however, the results acquired by the latter were preferred. Due to the anisotropic orientation of cardiac muscle fibers MLE would find non-existing data beyond the peak leading to a flatter and hence, falsified distribution. Therefore, Least Squares Fitting was used in order to gain reasonable results.

The fiber orientations illustrated in the intensity plot Fig. 2.10 confirm the visual impression gained by considering the 3D reconstruction of the image stacks obtained by means of volume rendering. The cardiac muscle fibers show high anisotropy as indicated in the intensity plot (e.g. Fig. 2.10), the preferred orientations appearing in red enclosed by dark blue areas. Bright blue or turquoise streaks, however, indicate greater isotropy as the fiber alignment is less expressed, but the dispersion increased. An example for an diminished fiber alignment is the intensity plot Fig. 3.18, where a large turquoise region is visible beyond a depth of 550 µm. However, as the SHG intensity decreases with increasing depth reaching the end of an image stack the quality of images is poorer, which might be the reason for the high dispersion in this case. High isotropy can also be observed in sample XXII slice 4 (see Fig. 3.20), where a fat deposit invades the cardiac muscle fibers. The analysis of the behaviour of relative amplitudes among the respective depth of the samples allows depth-resolved changes in the structural organization of the cardiac muscle fibers to be identified. While most samples showed a continuous orientation of preferred fiber directions (for example Fig. 3.8), some (see Fig. 3.12) did not have a conspicuous principal fiber direction as the change in orientation slightly varies among the depth. The location parameter μ plotted over the respective depth also illustrates the alternating orientations. This diminishes the overall rotation per depth, for example sample VIII b reaches a rotation of only 10.84° per mm depth compared to the other samples.

Analysis of 29 biaxial and triaxial heart samples results in an average rotation of 19.62 ± 7.32° per mm depth. By omitting the outlier (sample III a exhibits a rotation of 41.52° over one mm depth), the average rotation would not be strongly modified to 18.84 ± 6.10°. These results are close to the values reported in previous studies. Histological studies have shown that the orientation of fiber angles vary from roughly +60° to -60° across the ventricle wall (Rohmer et al. , 2007). Assuming the ventricle wall to be about 10−15 mm thick, the obtained results are consistent with published findings. Moreover, the total fiber rotation among the wall from epi- to endocardium has been reported to be 140° in rat (Pope et al. , 2008), 135° in guinea pig (Smith et al. , 2008), 140° in dog (Streeter et al. , 1969), 180° in pig, and 106° in mouse (Smith et al. , 2008). Small differences may arise due to distinct species, however, the porcine and the human myocardium are close to identical.

In order to determine the fiber rotation throughout the whole depth of the left ventricle wall, it was divided into consecutive slices that were imaged separately, but all at the same location. The rotation of each slice was determined, all together representing the rotation throughout the whole ventricle wall. However, enormous fat deposits (also small fissures or blood vessels are possible) are embedded in the microstructure, impairing the results. Due to the deposit, the rotation of slice 4 (medial) yields exceeding 192.11° over a depth of 800 μm, which adulterates the expedient determination of muscle fiber rotation throughout the wall. The numerous fat deposits, which are even visible to the naked eye (see Fig. 2.5), hinder a reasonable determination of fiber orientation and impede the detection of the same spot in each slice devoid of any deposits throughout the ventricle wall. However, factors such as age and weight of the patient as well as possible diseases might impact the continuous fiber structure and therefore, the results. Slice 3 of sample XXII (medial origin but nearby the epicardium) exhibits a rotation of only 3.95° over a depth of 980 μm. The reason for this minor change in orientation remains unclear. The need to analyze further samples in order to obtain reasonable results is necessary to gain more specific information about the fiber rotation across the whole ventricle wall. In contrary to previous studies that reported the muscle fiber direction to rotate continuously throughout the wall, this could not be demonstrated with the achieved results. Although this method is more labor-intensive, as it is a challenge to image exactly the same spot of each slice and does not avoid deconstructing the ventricle wall, this method offers a substantially better resolution than DTI. Moreover, the imaging itself is considerably quick. Cutting the slices by means of a microtome that would allow slices up to 1.5 mm thick, would improve the smooth surfaces of the samples that are necessary for intact imaging. Moreover, it is inevitable to store the individual samples fixed between, for example, two specimen holders to avoid possible deformations. These possible deformations would disturb an efficient imaging process, as the achieved penetration depths are diminished if there is space between the sample and the slide during the actual imaging process.

To show that the cardiac muscle fibers are organized in sheets, rectangular pieces of two heart samples were cut and rotated 90° around the x-axis prior to imaging. The images obtained proved that little to no fibers run in radial (depth-) direction. The calculated rotation of fiber direction is essentially smaller than recorded in xy-plane with the 3-D reconstructions illustrating the organization of fibers in sheets. The out-of-plane rotation was only determined for two specimen. However, these two samples both exhibited a similar rotation resulting in a mean radial rotation of 2.49°.

The left ventricle has to support a higher pressure than the right ventricle and thus, the wall thickness of the left ventricle is larger than that of the right one (Holzapfel & Ogden, 2009). Although the right ventricle wall is smaller, the obtained results show that the rotation across the wall is comparatively high.

The mean rotation per mm depth of the right ventricle is 45.6°. Assuming the right ventricle wall to be 4 mm thick, this would lead to a change of rotation of roughly 180° throughout the wall. According to our findings, the rotation throughout the right ventricle wall is approximately the same as throughout the left ventricle wall, the latter being about 10−15 mm thick. This leads to the assumption that the muscle fiber direction rotates faster through the right ventricle. The right ventricle has similar functions to the left ventricle, and therefore the overall rotation is approximately the same. However, the small quantity of only two samples of the right ventricle used does not allow for any conclusions to be drawn.

Considering all samples tested, it is not possible to extract additional information, neither visually from the 3D reconstructions nor from the determined fiber rotations, regarding the change in orientation referring to the location (either epicardial, medial or endocardial) of the sample in the heart. The fiber angles vary smoothly across the ventricle wall, but the variance in direction becomes greater close to the endo- and epicardial walls according to Rohmer et al. (2007). Also Streeter et al. (1969) reported that the greatest change in angle occurs close to the epi- and endocardial walls with respect to the ventricle wall thickness. However, this could not be explicitly derived from our obtained results. One demonstrative fact is that considering heart no. 5, the rotation is remarkably diverse close to endo- and epicardium. While sample III a (epicardial) exhibits a rotation of 41.52° per mm depth, the rotations of samples VII a, VII b, and VIII a (endocardial) are approximately half of that (sample VII a has a rotation of 16.13°, sample VIII a has a rotation of 17.28°, and sample VIII b has an even smaller rotation of 10.84° per mm depth).

Detailed information on how different cardiomyopathies have an impact on the 3-D fiber organization is not available. Therefore, the resulting data, such as rotation per depth and dispersion parameter have to be linked to the patient data. However, by the time this thesis was accomplished, we were not endowed with specific information about all the different hearts of which samples were derived.

The technique presented here has important implications for the knowledge about the microstructure of the human heart. DTI, which does not involve physical sectioning of the heart for the determination of the fiber rotation across the wall, offers similar findings. However, one essential convenience of the used method is the determination of material parameters. Deconstructing the heart wall can not be precluded as a maximum depth of 2 mm can be reached using multi-photon microscopy in combination with optical tissue clearing. The method used, however, gives more detailed information about the myocardial microstructure due to its improved resolution and enables the visualization of the fiber alignment in sheets. Moreover, it offers the identification of structural parameters and hence, the dispersion. The dispersion parameter that can be directly used in numerical modelling such as finite element method analyses was determined in human hearts for the very first time.

Bibliography

Campagnola, P. 2011. Second harmonic generation imaging microscopy: applications to diseases diagnostics. *Anal Chem*, **83**(9), 3224–3231.

Campagnola, PJ, Millard, AC, Terasaki, M, Hoppe, PE, Malone, CJ, & Mohler, WA. 2002. Three-dimensional high-resolution second-harmonic generation imaging of endogenous structural proteins in biological tissues. *Biophys J*, **82**(1), 493–508.

Chen, J, Song, S-K, Liu, W, McLean, M, Allen, JS, Tan, J, Wickline, SA, & Yu, X. 2003. Remodeling of cardiac fiber structure after infarction in rats quantified with diffusion tensor MRI. *Am J Physiol Heart Circ Physiol*, **285**(3), 946–954.

Cheng, A, Langer, F, Rodriguez, F, Criscione, JC, Daughters, GT, Miller, DC, Ingels, NB, John, C, & Neil, B. 2005. Transmural sheet strains in the lateral wall of the ovine left ventricle. *APS*, **289**(3), 1234–1241.

Cicchi, R, Vogler, N, Kapsokalyvas, D, Dietzek, B, Popp, J, & Pavone, FS. 2013. From molecular structure to tissue architecture: collagen organization probed by SHG microscopy. *J Biophotonics*, **6**(2), 129–142.

Cox, G, & Kable, E. 2006. Second-harmonic imaging of collagen. *Methods Mol Biol*, **319**, 15–35.

Cox, G, Kable, E, Jones, A, Fraser, I, Manconi, F, & Gorrell, MD. 2003. 3-Dimensional Imaging of Collagen Using Second Harmonic Generation. *J Struct Biol*, **141**(1), 53–62.

Eggen, MD, Swingen, CM, & Iaizzo, PA. 2012. Ex vivo diffusion tensor MRI of human hearts: Relative effects of specimen decomposition. *Magn Reson Med*, **67**(6), 1703–1709.

Erasmus, LJ, Hurter, D, & Naudé, M. 2004. A short overview of MRI artefacts. *S Afr J Radiol*, **8**(2), 13–17.

Fedak, PWM, Verma, S, Weisel, RD, & Li, R-K. 2005. Cardiac remodeling and failure. From molecules to man (Part II). *Cardiovasc Pathol*, **14**(2), 49–60.

Fenton, F, & Karma, A. 1998. Vortex dynamics in three-dimensional continuous myocardium with fiber rotation: Filament instability and fibrillation. *Chaos*, **8**(1), 20–47.

Gasser, C, Ogden, R, & Holzapfel, GA. 2006. Hyperelastic modelling of arterial layers with distributed collagen fibre orientations. *J R Soc Interface*, **3**(6), 15–35.

Han, M, Giese, G, & Bille, J. 2005. Second harmonic generation imaging of collagen fibrils in cornea and sclera. *Opt Express*, **13**(15), 5791–5797.

Hirshburg, JM, Ravikumar, KM, Hwang, W, & Yeh, AT. 2013. Molecular basis for optical clearing of collagenous tissues. *J Biomed Opt*, **15**(5).

Ho, SY. 2009. Anatomy and myoarchitecture of the left ventricular wall in normal and in disease. *Eur J Echocardiogr*, **10**(8), 3–7.

Holzapfel, GA, & Ogden, R. 2009. Constitutive modelling of passive myocardium: a structurally based framework for material characterization. *Philos Trans A Math Phys Eng Sci*, **367**(1902), 3445–3475.

Kumfer, K. 2009. Second harmonic imaging microscopy.

Kung, GL, Nguyen, TC, Itoh, A, Skare, S, Ingels, NB, Miller, DC, & Ennis, DB. 2011. The presence of two local myocardial sheet populations confirmed by diffusion tensor MRI and histological validation. *J Magn Reson Imaging*, **34**(5), 1080–1091.

Larson, AM. 2011. Multiphoton microscopy. *Nat Photonics*, **5**.

Ohayon, J, & Chadwick, RS. 1988. Effects of collagen microstructure on the mechanics of the left ventricle. *Biophy J*, **54**(6), 1077–1088.

Peeters, THJM, Vilanova, A, Strijkers, GJ, & Romeny, BMH. 2006. Visualization of the fibrous structure of the heart. *Pages 309–316 of: Vision, Modeling and Visualization 2006*.

Pope, AJ, Sands, GB, Smaill, BH, & LeGrice, IJ. 2008. Three-dimensional transmural organization of perimysial collagen in the heart. *Am J Physiol Heart Circ Physiol*, **295**(3), 1243–1252.

Punske, BB, Taccardi, B, Steadman, B, Ershler, PR, England, A, Valencik, ML, McDonald, JA, & Litwin, SE. 2005. Effect of fiber orientation on propagation: electrical mapping of genetically altered mouse hearts. *J Electrocardiol*, **38**, 40–44.

Ravichandran, R, Venugopal, JR, Sundarrajan, S, Mukherjee, S, & Ramakrishna, S. 2012. Minimally invasive cell-seeded biomaterial systems for injectable/epicardial implantation in ischemic heart disease. *Int J Nanomedicine*, **7**(Jan.), 5969–5994.

Rohmer, D, Sitek, A, & Gullberg, GT. 2006. Reconstruction and visualization of fiber and sheet structure with regularized tensor diffusion MRI in the human heart.

Rohmer, D, Sitek, A, & Gullberg, GT. 2007. Reconstruction and visualization of fiber and laminar structure in the normal human heart from ex vivo DTMRI data. *Invest Radiol*, **42**(11), 777–789.

Schriefl, AJ, Reinisch, AJ, Sankaran, S, Pierce, DM, & Holzapfel, GA. 2012. Quantitative assessment of collagen fibre orientations from two-dimensional images of soft biological tissues. *J R Soc Interface*, **9**(76), 3081–3093.

Schriefl, AJ, Wolinski, H, Regitnig, P, Kohlwein, SD, & Holzapfel, GA. 2013. An automated approach for three-dimensional quantification of fibrillar structures in optically cleared soft biological tissues. *J R Soc Interface*, **10**(Mar.).

Smith, RM, Matiukas, A, Zemlin, CW, & Pertsov, M. 2008. Nondestructive optical determination of fiber organization in intact myocardial wall. *Microsc Res Tech*, **71**(7), 510–516.

Sosnovik, DE, Wang, R, Dai, G, Reese, TG, & Wedeen, VJ. 2009. Diffusion MR tractography of the heart. *J Cardiovasc Magn Reson*, **11**(47).

Streeter, DD, Spotnitz, HM, Patel, DP, Ross, J, & Sonnenblick, EH. 1969. Fiber Orientation in the Canine Left Ventricle during Diastole and Systole. *Circ. Res.*, **24**(3), 339–347.

Williams, RM, Zipfel, WR, & Webb, WW. 2005. Interpreting second-harmonic generation images of collagen I fibrils. *Biophys J*, **88**, 1377–1386.

Printed in the United States
By Bookmasters